T5-BCA-514

Simplicius
On Aristotle's *Physics 2*

Simplicius
On Aristotle's *Physics 2*

Translated by Barrie Fleet

Cornell University Press
Ithaca, New York

For information address Cornell University Press,
Sage House, 512 East State Street, Ithaca, New York 14851

First published 1997 by Cornell University Press.

ISBN 0-8014-3283-9

Acknowledgments

The present translations have been made possible by generous and imaginative funding from the following sources: the National Endowment for the Humanities, Division of Research Programs, an independent federal agency of the USA; the Leverhulme Trust; the British Academy; the Jowett Copyright Trustees; the Royal Society (UK); Centro Internazionale A. Beltrame di Storia dello Spazio e del Tempo (Padua); Mario Mignucci; Liverpool University; the Leventis Foundation; the Humanities Research Board of the British Academy; the Esmée Fairbairn Charitable Trust. The editor wishes to thank Dr Margaret Atkins, Dr Gillian Clark, Dr Mark Edwards, Pamela Huby, Mr Edward Hussey, Dr Katerina Ierodiakonou and Professor Donald Russell for their comments, and Drs Rob Wisnovsky and Sylvia Berryman for their help in preparing the volume for press.

Printed in Great Britain

Librarians: Library of Congress Cataloging-
in-Publication data are available.

Contents

Introduction

Part One

Richard Sorabji

Book 2 of the *Physics* is arguably the best introduction to Aristotle. It contains ideas that are central to his thought, but also of continuing philosophical importance today.

In Chapter One, he defines nature, because his subject is natural science, and distinguishes natural objects from artefacts.

In Chapter Two, he distinguishes the subject matter of the natural scientist from that of the mathematician, although he relates the two.

In Chapter Three, he introduces his seminal distinction of the four causes, or four modes of explanation.

In Chapters Four, Five and Six, he explains what both luck and chance are: various kinds of coincidences. He does not yet make the anti-determinist decision, which I believe he later makes in *Metaph.* 6.3, that coincidences lack a cause, since they lack an explanation.[1]

After resuming in Chapter Seven the theory of four causes, he argues in Chapter Eight that there is purpose in nature, even in the absence of consciousness. A rival theory of purposeless natural selection can safely be rejected because it lacks the refinements of the modern theories of natural selection.[2]

In Chapter Nine, Aristotle shows how matter or the material cause explains: not as materialists think as a necessitating cause, but as a prerequisite presupposed for the attainment of natural purposes.

What does Simplicius add to Aristotle's bold theory of nature?

Simplicius cites the interpretations of many predecessors, reserving a special rivalry for the greatest commentator of the former Aristotelian School, Alexander of Aphrodisias. Alexander's commentary on the *Physics* is lost, except for some newly discovered excerpts from the later books, currently being edited by Marwan Rashed, who has used them to argue that Simplicius' reports of Alexander are unfair to him.[3]

[1] Sorabji (1980) ch. 1.
[2] Sorabji (1980) chs 10-11.
[3] Rashed (1996).

Simplicius also provides some very useful summaries of five distinct definitions of nature (*CAG* pp. 282,30-285,12), of Aristotle's first two chapters (309,2-31), and of his account of luck and chance (356,31-358,4).

Discussing nature, Simplicius sees a problem about the relation of soul to nature. Aristotle might arguably have been willing to treat the souls of living things as one kind of nature,[4] and Alexander takes this to have been Aristotle's view at least for the case of the supposedly living heavens.[5] But the Stoics disagreed, concentrating on the case of plants. Long before Descartes,[6] they rejected Aristotle's recognition of a non-conscious soul in plants with merely nutritive functions. Instead they substituted nature as the property of plants and contrasted soul.[7] This may help to explain why Simplicius and Philoponus as reported by Simplicius[8] can take nature and soul to be distinct agencies in living things. Despite this, they think that living things can be acted on by both soul and nature, but they also believe this calls for explanation, and Philoponus complains that Aristotle is not consistent in explaining celestial motion by soul as well as nature. Elsewhere, Philoponus identifies soul and nature in living things.[9]

If soul and nature are distinct, what is the difference? Simplicius says that nature is a principle that permits things to be moved passively rather than one that causes motion.[10] Aristotle had needed this idea in order to pave the way for his divine unmoved mover. For this he uses the rule that whatever moves is moved *by* something, and by something sufficiently distinct from itself. But how does this rule apply to things which move in accordance with their own inner nature, like falling rocks and rising steam? Their inner nature is not sufficiently distinct from them for Aristotle's purposes. He therefore insists that their motion requires a releaser, which acts as an accidental cause: the person who dislodges the rock, or takes the lid off the kettle, or boils the water in the first place. The inner nature of the rock or steam thus permits it to be moved passively by the releaser or generator. But does this distinguish nature from soul? For in the same book Aristotle says that the soul of an animal also has to be stimulated by the outer environment.[11] But he there also insists that

[4] Sorabji (1988) 222.

[5] ap. Simplicium *in Cael.* 380,29-381,2; 387,12-19; *in Phys.* 1219,1-7.

[6] Descartes, 'Reply to objections brought against the *Second Meditation*', §4, in the fifth objection, translated by Haldane and Ross, vol. 2, p. 210.

[7] Galen, *PHP* 6.3.7 (SVF 2.710); Clement, *Stromateis* 2.20 (SVF 2.714).

[8] Simplicius below *in Phys.* 262,13-263,11; 286,20-287,25; 379,28-9; also *in Cael.* 387,12-19. Philoponus, *contra Aristotelem* bk. 2, fr. 49-50, translated in this series from Simplicius, *in Cael.* 78,12-79,14; 199,27-35 Wildberg (1988) 160-5.

[9] Philoponus, *in Phys.* 2 197,4-5.13-22; 198,7 Lacey (1993).

[10] Simplicius below *in Phys.* 287,10; also *in Cael.* 387,12-19.

[11] *Phys.* 8.2, 253a7-20; 8.6, 259b1-20; cf. *DA* 3.10, 433b13-19.

in a certain sense soul is unmoved, or is moved only in a restricted sense.[12] This makes it different from nature.

In discussing the natural scientist as against the mathematician, Simplicius quotes a precious summary, at page length, of Posidonius' lost treatise on Meteorology, on the subject of mathematical astronomy.[13]

On the subject of causes, Simplicius gives the Neoplatonist list which expands Aristotle's four causes to six.[14] There is the instrumental cause[15] and the paradigmatic cause.[16] Aristotle had called his form or formal cause a paradigm, but he did not accept the Platonic Forms, which are what constitute the Neoplatonist paradigmatic cause.

Aristotle's efficient cause is rarely called a *poiêtikon* cause by Aristotle himself, but by the later commentators. Moreover, the late Neoplatonist commentators added a twist. Simplicius tells us of a whole book written by his teacher Ammonius, to show that Aristotle's God was not only a final cause of motion, but also an efficient cause of existence for the universe.[17] A *poiêtikon* cause is here a sustaining cause of being, not merely an efficient cause of other effects.

Simplicius reflects another Neoplatonist view about causation when he allows that causes need not be like their effects, in those cases where they are greater than the effects.[18] He is talking about the Platonic Forms as causes. An even clearer example in Plotinus is that of the One which is beyond the Forms.[19]

Simplicius is also well versed in the Stoic sub-distinctions among causes.[20] The Stoics took from Plato[21] the idea of joint causes (*sunaitia*). These used each other in order to produce the effect. Co-operating causes (*sunerga*) are defined by the Stoics as intensifying the effects of other causes or making them easier to achieve. They are not, like *sunaitia*, on a more or less equal footing with each other. The Stoic containing or cohesive cause (*sunektikon*) is the *pneuma*, i.e. the elements of fire and air which sustain things by permeating them and holding them together.[22] Containing causes are also sufficient causes which explain the behaviour of the things they hold together. They

[12] *Phys.* 8.5, 258a7; a19; *DA* 1.3, 406a3; b7-8; 1.4, 408b5-18; 2.5, 417a31-b16; 418a1-3. See Sorabji (1988) ch. 13.

[13] Below *in Phys.* 291,21-292,31.

[14] ibid. 316,23-6.

[15] ibid. 316,9.10.25; 317,24; 318,24.

[16] ibid. 298,17; 316,24; 317,31; cf. p. 363.

[17] Simplicius, *in Phys.* 1361,11-1363,12, translated with discussion in Sorabji (1988) ch. 15.

[18] Below *in Phys.* 297.

[19] Plotinus 6.7.17; Sorabji (1983) 315-16.

[20] Below *in Phys.* 316,25; 326,15-16; 359,18-21; 360,16; 370,15. For the Stoic sub-distinctions, see Frede (1980) 217-49.

[21] Plato, *Timaeus* 46C.

[22] Sorabji (1988) 85-9.

need to be triggered by prior or antecedent causes (*proêgoumena*). Simplicius does refer to nature acting as something prior (*proêgoumenôs*),[23] but if he means to be talking of the Stoic prior or antecedent causes, he has altered them. For the Stoics would think of a thing's nature as a cohesive cause. Simplicius refers also to containing causes.[24] He may be thinking of a containing cause in a non-Stoic way as one which incorporates others,[25] but he may have in mind the Stoic idea that it most fully accounts for behaviour. Simplicius also introduces a causal relation of his own: in some cases luck is responsible for causing other causes to achieve their end.[26]

In the section on luck and chance, Simplicius attacks both those who ascribe nothing to chance and those who ascribe too much. The latter are attacked also in his commentary on Epictetus as leaving no room for what is up to us.[27]

There is a discussion of missing a great evil or a great good by a hairsbreadth (*para mikron*). We think the former a godsend and the latter a disaster, for, as Aristotle says, 'the mind proclaims them [the great evil or good] already yours' (197a29). It was a Stoic view that things are often judged good or bad and give rise to emotion only because they are unexpected. We should counter this by expecting loss. Epictetus makes this point in his *Manual*, 3 and 21, on which Simplicius commented. Nonetheless, Simplicius does not here make the Stoic point that thinking a narrow miss a godsend or a disaster is purely a matter of expectation and so irrational.

In the section on purpose in nature, Simplicius refers to the old debate on whether the clever behaviour of animals is due to intellect and reason, to mere instinct or nature, or to something intermediate: a natural self-awareness which falls short of intellect and reason.[28] Intellect and reason were denied to animals by Aristotle and the Stoics, but ascribed by some Platonists, most notably by Plutarch and by Porphyry in *On Abstinence from Animal Food*, which is to be translated in this series.[29] Porphyry's view might seem to follow from Plato's idea that human souls can be reincarnated in animal bodies. Accordingly, Plato describes some animals as possessing reason, even if it is disused.[30] I believe, despite conflicting evidence, that Porphyry did at least entertain Plato's idea of the transmigration of human souls into animals. Conflicts in the evidence, I think, are due to

[23] 370,15.
[24] 326,15-16.
[25] See note 243 to the translation.
[26] 360,15f.: see note 343 to the translation.
[27] Simplicius *in Epicteti Enchiridion 1*, lines 175-97 (Hadot).
[28] Sorabji (1993) ch. 7; Simplicius below, 378,27-379,22.
[29] cf. Sorabji (1993) chs 1-7.
[30] Plato, *Timaeus*, 91D-92C.

Porphyry's following Plotinus' hesitations about whether the human soul is 'present without being present' to the animal, or whether it does not enter the animal at all. Later Neoplatonists, however, felt obliged to maintain against Porphyry the pagan practice of animal sacrifice. I suspect it was for this reason that they either took the idea of transmigration as a metaphor, or postulated that human rational souls guided animals only by remote control, without entering them.[31] In either case, the late Neoplatonists, including the author of the *de Anima* commentary ascribed to Simplicius,[32] rejected Porphyry's forthright ascription of reason to animals,[33] and this is why Simplicius speaks so cautiously when he says, 'In case anyone should think that irrational animals act because of any power of reasoning (*logismos*)'.[34] On the other hand, Proclus, distinguishing intellect (*nous*) from reason (*logos*) had been willing to ascribe intellect to some animals, not out of any respect for animals, but in order to expand the empire of reason.[35] That is probably why Simplicius seems so much more comfortable in this passage with the ascription to animals of *nous*.

I think it can be seen that the centuries of commentary on Aristotle and the development of Stoicism and Neoplatonism have enabled Simplicius to add much that is new to the reading of Aristotle.

Part Two

Barrie Fleet

Simplicius was a Platonist. He was born in Cilicia and studied in Alexandria before moving to Athens towards the beginning of the sixth century A.D. In Athens he was a leading figure in the Academy under its head Damascius until the closure of the pagan schools in 529 by the Christian Emperor Justinian. Whether Simplicius eventually returned to Athens, or whether he stayed in the East at Harran (Carrhae – earlier the scene of the disastrous defeat of Crassus' Roman legions in 53 B.C.),[36] he was free of the constraints of formal teaching and able to devote his time to writing. Consequently his commentary on Aristotle's *Physics* takes the form of an extensive and free-ranging treatise, in contrast to that of his contemporary Phi-

[31] Sorabji (1993) 188-94.
[32] Simplicius [?], *in DA 3* 187,35ff; cf. 211,1ff.
[33] Porphyry, *On Abstinence*, esp. bk. 3.
[34] Below, 379,15-16.
[35] Proclus, *in Tim.*, vol. 3, 330 (Diehl); *Elements of Theology* 64; *De Malorum Subsistentia* 25 (Isaac).
[36] See Appendix: The Commentators, 199.

loponus, the sections of whose commentary[37] have been plausibly identified by Etienne Evrard[38] as hour-long lectures given in Alexandria.

There was a long tradition within Platonism of seeking to harmonise the doctrines of Aristotle with those of Plato. Aristotle had developed philosophical thinking to such a degree that he could neither be ignored nor opposed on every question. Much of the work of harmonisation had been done in the centuries before Simplicius – Plotinus continued in the steps of Ammonius Saccas, by whom he had been taught in the early third century A.D. and who was concerned, according to Hierocles,[39] to find agreement (*homodoxia*) between Plato and Aristotle.[40] In late antiquity a thoroughgoing study of Aristotle was the foundation for a philosophical education, beginning with the logical works known collectively as the *Organon* (*Categories*, *On Interpretation*, *Prior* and *Posterior Analytics* and *Topics*), and continuing, *inter alia*, with the *Physics*.[41] Simplicius himself quotes widely from the works of Aristotle in his commentary, notably *On the Soul*, *Metaphysics*, *Categories*, *On the Heavens*, *On Coming-to-be and Passing Away*, *Topics*, *Nicomachean Ethics*, *Parts of Animals*, *Prior* and *Posterior Analytics*.

There remained, however, fundamental differences between Platonists and Aristotelians. Perhaps the most significant of these was the outcome of the basic divergence between the rationalism of the former, whereby a higher ontological status was accorded to the Intelligible World than to the Sensible World, and the empiricism of the latter, whereby this status was reversed. On the face of it, then, it might seem that *Physics 2*, which deals with just the sort of change in the Sensible World, the Being of which the Platonist at his most extreme would challenge, was likely to be an area of confrontation between the two schools. In fact it is, in the main, unproblematic. Far from altogether denying Being to the Sensible World, Plato's ontology does allow it 'Being of a sort'. In one of the dialogues which was most influential throughout antiquity, the *Timaeus*, Plato looks for explanations of change in the Sensible World, and the Neoplatonist successors of Plotinus find no difficulty in maintaining at one and the same time a Platonic ontology and what is essentially an Aristotelian explanation of nature.

Richard Sorabji has pointed above to instances in Simplicius' commentary where there are clear Neoplatonic influences at work,

[37] For Philoponus' commentary see Lacey (1993) in this series.
[38] Edwards (1994) 5.
[39] Hierocles *On Providence* in Photius *Bibliotheca* codex 214 (*Patrologia Graeca* 103.705d).
[40] See Appendix: The Commentators, 195-8.
[41] See Appendix: The Commentators, 194-5.

although by and large these do not fundamentally challenge or contradict anything that Aristotle says. But there are places where Simplicius shows himself to be unequivocally a Platonist, and it is there that we see most clearly the influence of Plotinus. Simplicius' prose on these occasions rises above its usually straightforward, sometimes tortuous, character. The most notable example is at 289,21-35, where the language is strongly reminiscent of Plotinus, and the tone elevated and religious. Noteworthy too is his talk of 'up there' (*ekei*) and 'down here' (*enthade*) to represent the Intelligible and the Sensible Worlds respectively – terms familiar both in Plato and Plotinus. But even on such occasions Simplicius is not seeking confrontation, and his Platonism sits happily alongside the material he is dealing with.

Simplicius lived and wrote nearly a millennium after Aristotle, but he is nevertheless very much a representative of the Ancient World, and there are times when his thought and his language are frustratingly obscure to a reader of our own times. But this is no more true of Simplicius than it is of Aristotle himself, and there are many shafts of light shed by Simplicius that illuminate the text as well as many modern scholars' attempts. A study of his commentary will, I hope, show an acute and original mind at work on material that is important but at times intractable.

I would like to thank Professor Richard Sorabji for his help and advice in the preparation of the translation and the notes.

Textual Emendations

277,26	*phusin* for *phusis*
291,22	add *tês* before *tôn*
292,14	*tropas* for *tropous*
300,4-5	move closure bracket from after *gnôrimôtera* (line 4) to after *eirêmenou* (line 5)
300,11	replace comma after *heteras* with question mark
322,11	*toû* for *tou*
335,19	I have omitted the phrase *dia tês tekhnês* (through art): see note 265
344,25	I retain *ouden*, bracketed out by Diels. See note 287
352,23	*elegon* for *elegen*
384,17	replace semicolon after *elthêi* with comma
386,5	add *zêtêsin* to fill lacuna indicated by Diels after *aition*

Simplicius

On Aristotle Physics 2

Translation

The commentary of Simplicius on Book 2 of the *Physics* of Aristotle

259,1

Aristotle has proposed in this treatise to expound the principles and causes[1] of what is constituted by nature. Of these principles some are elemental,[2] others are efficient and others final. The ones that most obviously present themselves to us are the elemental, because anything that can be known provokes the person wanting to know about it to enquire about its causes from its own nature and make-up, which the elements in it combine to produce. That is why Aristotle, in the first book of the *Physics*, after outlining and examining the doctrines of the natural scientists, proceeded to reveal the elemental principles; he showed that coming-to-be starts with the opposites (the most basic of which are form and privation) and also with what acts as substrate to the opposites. He has also demonstrated just what sort of a thing matter is: that it is a substrate for the opposites; that it is inherent in the compound body; by what species of knowledge it is to be apprehended by the natural philosopher; in what respect it differs from privation; that it is ungenerated and imperishable; and what other of the natural scientist's criteria can be applied to it. In just the same way he has demonstrated just what privation is: that it is absence of form in the naturally constituted object; that *per se* it is non-existent and exists only *per accidens*; that it does not function as an inherent cause, but only *per accidens* (by being absent); and that it is generated and is perishable.[3]

5

10

15

Having shown all this, he subsequently[4] intends to discuss what, in the case of the elemental formal principle, can be encompassed by the natural scientist, and then to move on to discuss the efficient and final causes. Since the form is inherent in the compound both as an element and as an efficient cause when interpreted according to its nature and its definition, and since he must explain about both the formal and the efficient cause, at the outset he quite reasonably discusses nature, which he will demonstrate as being both form and efficient cause.

20

260,1

Secondly, since some said that nature was matter, and others that it was form,[5] and since those who claim that it is form are the more accurate, he is bound, as regards the distinction between form and matter, to begin with an interpretation of nature.

5 Thirdly, if the matter which acts as substrate to natural objects bears the same interpretation as that which acts as substrate to what is artificial; and if the form which is concerned with the artificial has its source of change entirely outside the object while that to do with natural objects has it within them; then the person who intends to discuss the natural form must begin by distinguishing between the objects that are 'by nature' and those that are not.

The discussion of nature is proper to this enquiry for other reasons 10 too. For in the first book he was trying to find common principles of all change. So he admitted examples of change in the sphere of the artificial, suggesting 'the cultured' and 'the uncultured'; from this starting point, having distinguished what is by nature from what is not, he finally discusses the natural, ignoring the artificial. So of necessity he speaks of 'nature' and 'what is by nature' and 'what exists according to nature' by separating them in the first instance from 15 'what is not by nature'. In fact, at the end of Book 1 it had been claimed that he would speak subsequently about natural and perishable forms.[6] In general the person who writes about natural science must have an understanding of the terms 'nature', 'what is by nature', 'according to nature' and 'having a nature' – what each of these terms is and how they differ from each other. The understanding of nature is a prerequisite of all this. Therefore he explains it in its primary 20 sense; he does not think it worthwhile asking whether it exists,[7] since its existence is evident, a defence he himself will offer when he says more about it, after stating just what it is; rather he demonstrates what nature is – and part and parcel of this demonstration is the assumption that it does exist. He shows up the difference between the mathematician, the doctor and the so-called natural scientist, 25 since they all deal with what is controlled by nature; he clears up questions about the number of senses in which the causes are talked about, offering examples according to each signification.

Some say that luck (*tukhê*) and chance (*automaton*)[8] are causes – i.e. those who say that some things occur as a result of luck or chance. It is a consequence of what they say, even if they themselves fail to 30 realise this consequence, that by these two they are creating the two highest principles. For if they say that these are god and matter, how could it concur that the one acts while the other is acted upon unless there were some other factor responsible for the concurrence? And such a factor results from luck and chance, and these are not consequences, but prior conditions. Those who say that good and evil are principles will admit that separation into opposites is the result of 35 luck and chance, as is original local distribution of the opposites. So 261,1 Aristotle makes a distinction concerning luck and chance, systematically criticising his predecessors on the grounds that they had accepted them as causes but had said nothing about them. In this book he uses a clearer method of exposition, proving easier to follow both in the setting out of the subject matter and in his literary style.

<CHAPTER 1>

> **192b8** Of things that exist some exist by nature, while others have different causes for their existence. Animals and their parts, plants, simple bodies such as earth, fire air and water exist by nature. (For we say that these and suchlike exist by nature.) 5

Before everything else he both asks and demonstrates precisely what nature is;[9] for if nature is not properly understood it is impossible to understand what is meant by the terms 'by nature' or 'according to nature'; nor is it possible to understand any of the natural entities[10] in so far as they are natural.[11] He discovers precisely what nature is from the difference between natural and non-natural entities. Those 10 that owe their substantial existence[12] to nature[13] he says exist by nature, while those that owe it to other causes do not exist by nature. For there are many other causes of things that come into being[14] – for example intellect and reason, whether it be practical or productive;[15] when it is linked with desire (*orexis*) it occurs as choice (*proäiresis*),[16] and gives scope for virtue and vice as exemplified by just or unjust action, while when it is without desire its scope is that of art (*tekhnê*),[17] exemplified by a bed or a house or flute-playing. Certain 15 things occur also by luck,[18] such as the windfall of a large sum of money, while others again happen as a result of chance, such as the occurrence of a portent or the way a stone falls to form a seat.[19]

Taking it, then, as evident that some things occur through causes other than nature, he proceeds finally to add what does occur by nature, saying that what exist by nature are animals and plants and their parts, and the simple bodies.[20] For choice is not responsible for 20 them, since choice is indeterminate, while their generation is determinate.[21] Nor do they owe their being to luck or to chance, for such things are less often the case, while things that owe their existence to nature are most often the case. Nor are they the product of art, for the products of art owe their existence to something external, while the products of nature owe it to what is within themselves.[22]

It is clear that all these are said to exist by nature in like manner, in so much as they have something in common.[23] Therefore they are not alike in respect of possessing the faculty of perception (for this is 25 particular to animals) nor in respect of nourishment, growth and reproduction; but they are alike in what are specifically called natural changes, both changes of place that happen without any impulse from the soul, and changes such as qualitative alteration, coming into and passing out of being, increase and decrease.

Having included 'animals', he also adds 'and their parts', 'because', 30 as Alexander says,[24] 'the parts of things that exist by nature also exist by nature. For not all the parts of things produced by art are themselves products of art (for of a house – which does exist by art – some components such as bricks and doors are products of art, while others such as the stone and wood are products of nature), while the parts of natural things are themselves the products of nature. This 35 is quite plausible; for natural bodies are the raw material of art (that is why they too are the parts of things produced by art), and of things that come into being by nature the raw materials too are natural; therefore their parts are all natural.'

262,1 But perhaps one should pay close attention to what he has said. For if we mean parts in the strict sense,[25] then the parts of the products of art are themselves products of art; for example, the head and the feet of the statue are products of art, as are the parts of the house – the men's and the women's quarters and the colonnades. And 5 if we call the elements (*stoikheion*) parts, such as the wood, stone and suchlike in a house, even in the case of natural things the primary elements are not themselves natural; for not even the matter is natural, since it is *per se* without a principle of movement.[26]

But perhaps there is some distinction to be made even in this case. For in fact in the case of the products of art the proximate elements are natural and not themselves the products of art – for example, the bronze of a statue and the stone and timber of a house; on the other 10 hand, in the case of natural entities, even if the ultimate substrate is not natural – such as its matter – at least the proximate components are, such as the four elements. So it would be in this way that there is a difference between what does and what does not exist by nature.

But in what sense does he mean that animals and plants exist by and because of nature? For they are ensouled, and are what they are because of soul.[27] In fact, when we define an animal we say that it is 15 a being with soul and with sensation, and the plant has been given its psychic character by the vegetative soul which causes it to take nourishment, to grow and reproduce its like. What is surprising is that Aristotle himself, a little further on, says that things which change by increase and decay or diminution suffer this change in so far as they are natural things – yet in his work *On the Soul* he says that it is appropriate for the vegetative soul to nourish and to give 20 increase and diminution.[28] And even further on[29] he says that when pieces of wood are planted in the earth they put forth shoots because of their own nature, although the shoots and the whole process of their production belong to them because of soul. The definition given of the soul[30] which calls it 'the first actuality of a natural body with organs, potentially possessing life' fits both plants and animals, but not the simple bodies such as earth and fire etc. which do not have 25 organs. Why, then, at this point when he wants to discover the difference between things that do and things that do not exist by nature – since it is in the distinction that nature is to be found – does he include animals and plants among things that exist by nature – the very things that are characterised by soul?

Perhaps, then, only the simple bodies are natural; but even if animals and plants have soul as well – the former a desiderative one, 30 the latter a vegetative one – they also have the nature that was the subject of enquiry, the nature which also the simple bodies have. For in fact the more perfect bodies, which have more perfect life, have as well the less perfect life. For example, man has a rational, a desiderative and a vegetative life and the nature which is the object of enquiry; the irrational animal has all the other lives except the 263,1 rational; the plant has the vegetative life and nature; and the simple bodies and their compounds, in so far as they are just compound bodies such as stones, wood, bones and in general inanimate bodies, have only nature – whatever this nature is. That is bound to be the case, since the soul supervenes[31] on a natural body as 'the first 5 actuality of a natural body with organs, potentially possessing life', as he himself defined it in *On the Soul*.

Further, if nature belongs to the simple bodies not in so far as they have organs but because they are compounds of form and matter, and if these bodies are inherent in plants and animals – for they too are made up of the four elements – it is clear that the bodies of animals 10 and plants are in all respects natural prior to their being those of animals and plants. That is perhaps why he added 'and their parts'. For bone and wood can no longer be said to have desire or a vegetative soul, but they do have a nature. He pointed out the difference between them and the things produced by art, and he included them in that they are natural, but said that increase belongs to natural bodies either because he had not yet determined to what things increase is 15 proper or because it was agreed by some to belong to certain things which are strictly speaking natural, such as fire – that is why he now includes <increase> as something natural.[32]

Eudemus[33] shows that Aristotle included animals and plants among natural things not in so much as they are animals and plants, but in so much as they are natural in themselves; in the first book of his *Physics* he writes: 'Since we say that many things exist by nature 20 (viz. a horse, a man and every animal and their parts; olive and every form of plant life and their parts; grass and in general all things that grow; earth and fire and many lifeless things), what is it that belongs to all of these? Sensation and various other features are particular to animals, while increase is particular to living beings, but more or less

25 all things can move (viz. wood, bronze, fire and in general all body) although not all their movements are similar. For example, a stone and any heavy object can move upwards and sideways only by some agency, but can move downwards of its own accord; fire can move downwards only by some agency but upwards by itself; wood moves downwards of itself, but the bed moves downwards in so much as it is made of wood, not in so much as it is a bed. For if it grows wings, 30 it will not be borne downwards.' If indeed things produced by art do not move of themselves but in so far as they are made of such materials, while natural things do move of themselves; and if natural things move sometimes because of movements imparted by something else, but sometimes because of self-imparted movements; and if we say that movements imparted by something else are contrary to nature, while self-imparted movements are according to nature; and if being moved according to nature would mean having the cause 264,1 of movement within themselves and not from outside; if all this were the case, then surely we must admit that nature is such a source of movement, since it happens to belong to all things that exist according to nature. If this is so, then nature becomes the source of internal and self-imparted movement.

> **192b12** All these things are clearly different from what is not constituted by nature. All the things that do exist by nature clearly have within themselves the source of change and of its cessation – either in respect of place, or of growth and decay, or 5 of alteration.

He has proposed to discover just what nature is by systematically revealing the difference between what exists by nature and what does not exist by nature but through other causes; this difference he concludes to be nature. Things that exist by nature differ from those that do not exist by nature in no other way than that 'they have within themselves a source of change and cessation of change';[34] by source 10 he means the efficient cause.

Just as natural things clearly change from within themselves – some things 'in respect of place' (such as earth which moves downwards and fire which moves upwards), others 'in respect of growth and decay' (such as fire itself and any rocks that can be observed increasing and decreasing in their entire bulk[35] – which is growth and decay) and others 'in respect of alteration' (such as water when it is 15 heated and rarefied), in just the same way they have cessation of such change within themselves. For the change and its cessation do not originate from outside, nor are they without limit; rather the change proceeds as far as the limit of the appropriate form and then ceases.[36]

Alexander notes: 'He says that each of these has a source of change and cessation of change within itself, referring to the aforementioned things, *viz.* animals, plants and simple bodies, but not all natural things. For in fact a body which is rotating, which too is a natural body, has within itself a principle of change but not of cessation of change, since it moves never-endingly.' But perhaps the heavenly bodies, even if they do not change from motion to its cessation, do enjoy <a sort of> cessation around their centres, axes, poles and in their entirety. Furthermore, since their motion is natural, its cessation would also be natural. But not all cessation is rest; only that after movement is.[37] 20 25

Porphyry[38] notes rather more aptly: 'Perhaps in the phrase "some source and cause of motion and rest" the word "and" instead of "or" is used on account of the fact that some of the things that exist by nature are always in motion and never at rest. I do not mean,' he says, 'divine body (for that is set apart from generated things), but fire, which is never at rest, since it is always moving up and down or in a circle. But it has some of the nature of the compound body.'[39] 30

Natural entities, then, have the cause of change and its cessation within themselves and not from outside, as is the case with the products of art and of other causes. When Alexander[40] says that perhaps (a) the simple bodies have within themselves the source only of local change, (b) animals have the source of all types of change, and (c) plants have the source of all changes except local movement, we will judge either that he is there including not the change in animals *qua* animals or of plants *qua* plants, but *qua* natural bodies, or else that he is taking up the argument in generalised terms. 265,1 5

192b15 But a bed and a cloak and anything else of that kind, in so far as they fall under such a predicate and are the products of art, have no innate impulse to change – although in so far as they are made of wood or stone or some composite of these *per accidens*, they do have, in this respect, such an impulse. For nature is a source and cause of change and its cessation in those things in which it is present primarily, *per se* and not *per accidens*. What I mean by 'not *per accidens*' is this: A man who is a doctor might be the cause of his own good health, but he would not possess his medical knowledge in so far as he is in good health; it is rather the case that doctor and patient are one and the same person *per accidens* – that is why, on occasions, their roles are seen as separable. It is the same with other things that are produced; none of them has within itself the source of its own production. In some cases, such as a house or anything that is manufactured, this source resides in something else and is external, while in others it does reside in the subject, but not in the subject *per se*, i.e. when the cause is there *per accidens*.

Having indicated from clear evidence that natural things have a 10 source of change and its cessation within themselves, he shows, by contrast with things that do not exist by nature, that this is also particular to things that do exist by nature. Things that do not exist by nature, in so far as they are described in this way – things which exist by art such as a bed or a cloak (this is what is meant by the phrase 'in so far as they fall under such a predicate') – 'have no impulse of themselves to change'. He called this source of change from within an impulse in the proper sense of the word; but some people 15 write 'source' instead of 'impulse'.[41] For things that exist by nature have the cause of change within themselves, while things that exist by art have it from outside – since neither the craftsman nor the art is within them. The clod of earth does not fall downwards because it is moved from outside, while the bed is fashioned from outside itself. Similarly with cessation of movement and change: in the case of the clod it occurs because it gains its unity[42] from within itself, while in 20 the case of the fashioning of the bed it occurs from outside because of the craftsman. In sum, the source of change is the same as the source of its cessation for things that exist by nature; similarly for those that exist by art.

In fact things that change naturally arrive naturally at the cessation of change, those that change their location by coming to their proper place, those that increase by coming to their proper size, and those that change qualitatively by coming to their proper form.[43] They 25 do not change without limit; and the cause of the cessation of their change is not external but within them. On the other hand, the bed and the cloak and in general anything that is made by art, in so far as they are made by art, have the source of change and its cessation from outside;[44] but in so far as the substrate of each of these is natural body – wood or wool or some other simple body or a mixture of simple bodies, in this respect even these things have within them the source 30 of change and its cessation.

If, then, all natural things have this source within them, and if the 266,1 products of art do not have it in so far as they are products of art but do have it in so far as they too are natural, then to have within themselves a source of change would be the particular property of natural things *qua* natural. It is therefore clear that nature is nothing other than a source of change and its cessation 'in those things in which it is present in a primary manner, *per se* and not *per accidens*'.

5 It seems that the conclusion which has been drawn can be put syllogistically according to the first figure[45] as follows:

Nature is that by which things that exist by nature are differentiated from those that do not.
Things that exist by nature are differentiated from those that do not by having an internal source of change and its cessation in a primary sense, *per se*, not *per accidens*.
Therefore nature is a source of change and its cessation in those things in which it is present in a primary manner, *per se* and not *per accidens*.

It can be framed according to the third figure as follows: 10

Things that exist by nature differ from those that do not by having a nature.
Things that exist by nature differ from those that do not by having within themselves a source of change and its cessation *per se* and not *per accidens*.
Therefore things that have a nature have a source of change etc.
Therefore nature is a source of change *per se* and not *per accidens*.

The syllogism is of this sort:

All beings capable of laughter are men. 15
All beings capable of laughter are rational mortal animals.[46]

The inference that all men are rational mortal animals is drawn as a general implication because of the matter, although the form of the syllogism yields <only> particular conclusions. For the propositions which contain the particulars and the definitions specified are in general affirmative and are convertible in a general way on account of the particular quality of their matter. That is why in the third 20 figure, even if as a result of the figure they draw the conclusion in a particular manner, even so, because of the matter the general case is true as well, as Aristotle himself says in the *Analytics*.[47] But the proposition which states 'Things that exist by nature differ from those that do not in respect of a source and cause of movement and its cessation' would be clarified by the following words: 'All these things are clearly different from those that do not exist by nature' etc. As Alexander says, 'The argument can be stated hypothetically as a 25 consequence in a positive form as follows: If things that exist by nature and have a nature differ from those that do not exist by nature in that they have within themselves a source and cause of change *per se* and not *per accidens*, then nature would be a cause of change in those things in which it resides *per se* and not *per accidens*; but if the first <is true>, hence the second. That it is in this respect that the 30 things that have a nature differ from those that do not, he confirmed in his inductive reasoning by putting forward by way of contrast the products of art.'

It was entirely necessary to add the words 'in those things in which 267,1 it is present primarily, *per se* and not *per accidens*'. For in fact the bed has within itself a source of movement and its cessation. At least, when it is dropped it falls downwards – although it is not said to be something natural or to have a nature *qua* bed, since the source of movement does not belong primarily to the bed but to the wood, and only because of the wood does it belong to the bed too. So the addition 5 of the word 'primarily' was necessary. And Aristotle himself explained why it was necessary to include '*per se* and not *per accidens*'. For if a thing is to be natural, it must have within itself the source of change in so far as it is what it is;[48] for example, earth is said to have a nature because it has within itself a source of downward movement in so far as it is earth. On the other hand, if a doctor who is ill treats 10 himself, he is treated by himself (since he has within himself the source of change), but *per accidens*, not *per se*. For the doctor is treated by himself not *qua* doctor, but *qua* patient.[49] For he gives the treatment in so far as he is a doctor, but receives it in so far as he is a patient. It is one thing for him to be a doctor, another to be a patient. That is why their roles are seen as separable. For not every doctor is 15 ill, nor is every patient a doctor, so that the doctor who is being treated by himself does not have the source of change within himself in so far as he is a patient; for if this were the case, then every patient would treat himself. He is treated *qua* patient, but he does not have the source of change, viz. being treated by himself, *qua* patient.

He illuminated the phrase '*per accidens*' well by saying 'that is why 20 on occasions their roles are seen as separable'. For what belongs to anything *per se* is inseparable from it, while what is separable does not belong to it *per se*, but *per accidens*. 'Primarily' differs from '*per se*', and not everything that belongs to something else *per se* can be said to belong to it primarily, nor *vice versa*. For when something belongs to something else *per se*, and it in turn belongs to some third entity *per se*, then the first entity belongs to the third *per se* but not 25 primarily; for example, having the sum of its three angles add up to two right angles belongs to the triangle *per se*, and triangularity belongs to the isosceles *per se*. Therefore having its three angles add up to two right angles belongs to the isosceles *per se*. For *qua* isosceles it is a triangle, and *qua* triangle the sum of its three angles is two right angles, and its triangularity and having the sum of its angles add up to two right angles are inseparable from the isosceles. But 30 having its angles add up in this way does not belong to the isosceles primarily, but through the mediation of the triangle. Again, whiteness belongs to the visible surface primarily, and virtue belongs to the soul primarily, for there is no intermediate. But these do not belong *per se*. For they do not form an essential part of the being of their subjects, nor are they included in the definition. In addition, white is by nature separable from visible surface, and virtue from soul.[50] 268,1

But sometimes both[51] do coincide in the same entity, when they do form an essential part of its being and are there without any intermediate; for example, the power of rational thought could be said to belong primarily and *per se* to a man, as could having the sum of its angles add up to two right angles be said to so belong to the triangle, and triangularity to the isosceles. But if the ship has the helmsman within itself moving it, it does not have him *per se*. For he does not complete its nature, since any ship, just to be a ship, would have no need of someone outside to move it; its helmsman could otherwise not be removed. For none of the things that belong to anything *per se* can be separated from it if the thing itself is to remain what it is. And even if the helmsman, in moving the ship, moves himself, he does not do so either *per se* or primarily. We should not agree with Alexander when he says that the soul moves itself like a helmsman by moving the body in which it resides, as the helmsman moves the ship. For although one might perhaps agree that the local movement of the soul takes place in this way, nevertheless its desires, its processes of thought and opinion and its impulses all occur when it moves itself primarily (there is no intermediate) and *per se*, for its very essence is self motion.[52] 5 10 15

Nor should we accept the following argument, when he says 'We should note that he included the soul in his description of nature, since according to him the soul is the actuality of the natural body with organs, while nature strictly and primarily resides in simple, not organic, bodies'. Soul is the actuality of the natural organic body, but nature would not be the actuality of the natural body, and the former would be a source of some changes, and the latter of others. And since nature resides, according to Aristotle, in the substrate body (for that in which it resides primarily is called the substrate), and since not all soul resides in a substrate (for intellect is separable, as he himself proclaims in *On the Soul*[53]), even if Alexander wants to understand it not as the psychic but as the divine intellect,[54] he matches that argument with his own opinion of the soul in a most unconvincing and polemical manner. But we too agree that the soul is a source of change, and a more powerful source than nature,[55] and not only a source but a fountain-head, and not present in what is moved as in a substrate. Aristotle wants the term 'in a substrate' to be added to nature when he says 'Each thing is a substrate and nature is in a substrate'. This indicates that the soul has activities separable 20 25 30 269,1 from the body,[56] as Aristotle said in *On the Soul*, demonstrating that it also has its essence separable from the body.

5 He called things that come into being by art 'things made', since art 'is a disposition allied to true reasoning that makes things', as he himself defined it in the *Nichomachean Ethics*.[57] In Book 7 of the *Metaphysics* he says that coming into being is properly said of what comes into being by nature, and that other kinds of generation are productions.[58] All things that happen or are made by art, then, are either initiated from outside, or, if from inside, *per accidens*.[59] The 10 great Syrianus[60] notes: 'This given definition of nature will fit almost all natural phenomena when applied properly to each particular case. For just as the name "nature" is predicated homonymously of both matter and Form and in the case of natural bodies is properly applied to what results, as it were, from nature as from a cause, as it were, 15 so the definition is applied of itself in the case of what is properly called nature, and by analogy in the case of the other sources. For the other natures are sources of change, but not in the same way.'

> **192b32** Nature is thus what has been stated; whatever has this sort of a source has a nature, and all these things are substances, for each is a substrate, and nature always resides in a substrate. These things are according to nature, as is anything that belongs to them *per se*, as moving upwards belongs to fire. For that is not nature, and does not have a nature, but is by nature and according to nature. It has been stated what nature is, and what 'by nature' and 'according to nature' are.

20 Having defined nature and having articulated any obscurities in the definition, he adds: 'whatever has this sort of a source has a nature'. Yet having indicated the things that have a nature from their activity,[61] from these he concluded just what nature is. So why does he then add: 'whatever has this sort of a source has a nature' as if it followed that nature was this sort of a thing? Now he did not add this as a proof; but although it was included as being evident, it renders 25 him a service having now been assumed, either (a) with regard to distinguishing between 'having a nature' and 'according to nature', which he does next, or (b) with regard to his conclusion that every natural thing is a compound substance having that in which its nature resides as a substrate – since nature is a source and cause in the thing in which it resides *per se* – but also having its nature itself 30 as something in a substrate.[62] Thus he adds as being consistent with this the fact that some people say that nature is the substrate of the natural object which is a compound,[63] from which things that come to be do so as from something inherent, while others say that it is the form. He arbitrates between the two accounts. He distinguishes

between 'that which has a nature' and 'what is according to nature', saying that what has a nature is a substance compounded from the substrate which contains it and the nature in it. For that which has a nature comprises a substrate and the nature in it. All these things that have a nature are substances. For substance has in a way been said to be both the matter and the substrate in the first instance,[64] and substance has also been said to be what is in the substrate, namely the form. But substance in the more proper sense is the compound. That things which have a nature are substances is therefore shown by the fact that there is a substrate and that nature resides in it, and that substance is each of these or rather the compound. 270,1 5

Or the words 'and all these things are substances' should be understood to mean that what has a nature, viz. the substrate, is a substance, and the nature is a substance, and *a fortiori* the compound is a substance, since there is 'a substrate and something in a substrate'. Or else if he has said that all these things are substances, namely the substrate and what is in the substrate, he shows this by the mediation of nature, when he says 'nature is a substrate, and always resides in a substrate'. If, then, nature is a substance, and if the substrate is a nature, and if what is in the substrate, namely matter and form, are natures, then substance would be all these. The question pursued shortly after this, whether it is the matter and the substrate or the form and what is in the substrate which comprise nature, seems to be in keeping with this way of thinking. 10 15

The former interpretation is the better both because of the subsequent conclusions and because Aristotle would not call the substrate and what is in the substrate, which are just two things, 'all these things'. At any rate, in *On the Heavens*[65] he wrote the following: 'Our language too shows the same tendency, for of two things or people we say "both", not "all". This latter term we employ when there are three in question.'[66] Therefore I omit the rest of Alexander's explanations concerning this topic. 20

All things, then, that have a nature are substance because something containing a substrate and something contained in a substrate is what nature is, or – to say the same thing – matter and form, which comprise substance in the proper sense of the word. Alexander notes that Aristotle, although he wants them to be substances, says that both form and nature reside in a substrate, when he claims in the *Categories*[67] that there is no substantial being in a substrate. Alexander softens the objection by saying either (a) that of the substances mentioned there, none was mentioned as being in a substrate (for Aristotle did not construct his argument there about substance under the heading of form, but of the compound, to which he says there is no opposite) or (b) that Aristotle does not now mean what is properly 25 30 said to reside in a substrate, which he had in mind in the *Categories*, but that he now means that that which craves some substrate is now spoken of in a substrate, just as in many cases what is *per accidens* comes from a substrate.

35 Having said that the things that have a nature are like this, he adds further conclusions concerning what 'the things that are according to nature' are, saying that they cover a wider field than the things that have a nature. For in fact the things that have a nature are said to be according to nature as being characterised by the nature within 271,1 them and as being just what they are said to be. But it is not only these things that are said to be according to nature, but also 'anything that belongs to them *per se*'. Fire has a nature by having within itself a source and cause of upward movement, and upward movement belongs to fire by nature and according to nature. But it can not be 5 said that to rise upwards 'has a nature', nor is this a nature. Nor is it even a substance, but a power and an activity belonging to fire according to the description of its nature; in this way it is said to be according to nature, and the same characteristic is said to be by nature, since <fire> has such a power and such an activity because of its own nature and because it is naturally like this.[68]

To rise upwards belongs, then, to fire both by nature and according 10 to nature. But 'by nature' is not the same as 'according to nature', since the former term has a wider application than the latter. We say that the term 'according to nature' belongs to those of the things that exist by nature which have their appropriate completion. There are certain things that exist by nature because they have come into being according to the activity of nature but which are not, however, according to nature; for example, things that are defective from birth 15 and in general anything suffering a privation. For such things are like this because nature has made them like this, and the privation is a lack of something in what nature has made. So the phrase 'by nature' would be applied to everything that is in conformity and in keeping with the natural substance as such; for example, colour and sickness belong to the body by nature. But the term 'according to nature' can be applied only to that which occurs according to the desire of nature. So we say that being healthy is according to nature, 20 while illness is present by nature, since it is contrary to nature. So when Aristotle says that what exists by nature also exists according to nature, he does not mean that these two terms are synonymous, but that it is in the case of fire that rising upwards belongs to it by nature and according to nature.

193a3 It would be ridiculous to attempt to prove that nature exists. For it is obvious that many existing things are like this.

> But to try to prove the obvious from the unobvious is the mark of a man incapable of distinguishing between what is self-evident and what is not. However, it is clear that this can be the case. For the man blind from birth could only make logical inferences about colours. Consequently, such people can argue about names, but not have any knowledge.

Since we should ask whether the object of our enquiry exists at all (25) before we ask the question what it is, as Aristotle explains in Book 2 of his *Posterior Analytics*[69] (since it would be a waste of time to ask what the proposed object of enquiry is if it does not exist at all, because there could be no definition of what does not exist); since, then, this is the case, and since he himself gave the definition of nature without showing that it exists, he explains that it was with good reason that he omitted to discuss whether it exists. For those things that have (30) evident substantial existence need no proof of their existence, as Aristotle went on to show in Book 2 of the *Posterior Analytics*, if I remember correctly. Because it is obvious from plain fact, he says, that things which have a principle of change within themselves, i.e. nature, exist, it is ridiculous to try to show that nature exists. For first, the fact that the person wishing to prove the obvious is forced, (272,1) if he is to be persuasive, to have recourse to something less knowable than what he is trying to prove, is typical of someone who is ignorant of the method of proof, namely that 'all demonstrative explanation and understanding depend on pre-existing knowledge',[70] as Aristotle himself explains. Secondly, the words 'the mark of a man incapable of distinguishing between what is self-evident and what is not' typify (5) the man who is anxious to prove by means of other things that nature, which is self-evident, is not self-evident. And it is even worse if they are to be proved by means of what is less knowable, which is what must happen in the case of things that are all too obvious. The man who wants to employ proof for everything eventually destroys proof. For if the evident must be the starting point of proof, the man who thinks that the evident needs proof no longer agrees that anything is (10) evident, nor does he leave any basis of proof, and so he leaves no proof either.

He brings in the blind man who makes inferences about colours to witness the fact that 'to try to prove the obvious from the unobvious' happens to those who have lost the means of judging the obvious. For colours are obvious and credible in themselves to anyone who can see, (15) but the person who has lost the faculty of sight which distinguishes the obvious in colours, and who then proceeds to make inferences of reason about them, will employ the less knowable in order to lend credence to his words, since he will be ignorant of the obvious and will construct his argument about mere words without any real

20 understanding. He makes his inferences from what he hears, not his own consciousness, since he does not have the knowledge which is consonant with the senses, without which not even his powers of rational inference can understand anything about colours. In this way the man trying to prove the obvious is ignorant of the nature of what is knowable by the senses, because he does not have the faculty by which these things are judged. This faculty is intellect. For it is 25 intellect, he says, by which we come to know that the terms of propositions are credible in themselves, obvious and simple. Perhaps it is the intellect which discovers that definitions ought to be obvious, credible in themselves and knowable without any rational inference, since they are made as a result of a clear division. Objects of inquiry which are self-evident, e.g. that the good is beneficial and suchlike are, as it were, terms that do not need anything else to prove them, 30 as ambiguous propositions do. The man blind from birth would make a rational inference about colours from what he hears and not from his own consciousness, that gray is fit for vision, saying 'gray is composed of the penetrative quality of white and the compressive quality of black; as such it is fit for vision'.[71] But he has no conscious- 35 ness of what penetrates or what compresses the visual faculty, nor of anything that is fit for vision; rather he merely applies the words. So 273,1 we should realise from what Aristotle says that it was quite reasonable for him not to enquire whether nature exists, and for him to give the reason for the omission not at the outset but at this point, when he has demonstrated what it is as something obvious. For all bodies are naturally capable of change.

5 **193a9** Some consider nature and the essence of things that exist by nature to be that which is primarily inherent in each thing, unformed *per se*; for example, the nature of a bed is its wood, of a statue its bronze. Antiphon says that an indication of this is the fact that if you were to bury a bed and if the decay were to have the power to put forth a shoot, then it would be a tree that grew, not a bed, since the conventional construction and workmanship belong to it only *per accidens*, while its essence is that which persists while undergoing such affections. Similarly, if each of these materials stands in a similar relationship to something else, for example bronze and gold to water, or bones and wood to earth etc., then these elements will be their nature and their essence. That is why some say that fire is the nature of things – others say it is earth, others air, others water, others two or three of these, others all of them. For whatever element (or elements) any one of them claims fulfills this role, that is what he says is the whole essence of anything, while all else are

affections, states and dispositions of it. Any one of these elements is said to be imperishable, not having the ability to change out of itself, while everything else passes in and out of being ceaselessly.

He said earlier that things which have a nature are substances, because that which has a nature is a substrate, and the nature which is contained is in a substrate (or because the nature is a substrate and is in a substrate, as Alexander understood it according to one explanation); so he quite reasonably now cites the views of those who say that nature is the substrate and of those who say it is what is in the substrate, and he outlines and criticises the arguments of each of the two groups. At the same time he tells us the meanings of the word 'nature', which is spoken of in several ways, as he himself has made clear at the end of this passage where he says 'Since it has been determined in how many ways nature is spoken of ...'[72] – I shall comment on it *ad locum*. When he said 'Some consider nature', he added 'and the essence of things that exist by nature' because the being and essence of things that exist by nature are according to nature. 10 15

First, he speaks about those who think that the matter is the nature of each thing, i.e. 'that which is primarily inherent in each thing, unformed *per se*; for example, the nature of a bed is its wood, of a statue its bronze'. To begin with he said 'unformed'[73] because in each composite entity there are often several things which go by the description of substrates; for example, in the bodies of animals the organic parts act as substrates immediately below the whole form, while the homoeomerous[74] parts act as substrates to them, the so-called four elements as substrates to them, and primary matter to them – primary matter is that which is primarily lacking in form *per se*.[75] All the other things, such as the organic parts, the homoeomerous parts and the elements, are lacking in form in relation to something else (the form that is imposed on them), although they do have their own forms, while primary matter is unformed *per se*. For bronze and wood, the matter of a statue or a bed, have their own forms and are only lacking in form in relation to something else, but the matter which is common to all things is lacking in form *per se*. The bronze and the wood are analogous to the primary matter, for just as they stand in relation to the statue and the bed, so it stands in relation to all things that have their own forms. Therefore they say that 'that which is primarily inherent in each thing, unformed *per se*' is the primary and common nature in the sense that the proximate matter of a thing is its proximate nature, for example bronze of a statue and wood of a bed. 20 25 30

To show that nature is the substrate and not the form, Antiphon 35 the Sophist[76] adduced the fact that it is nature which either causes things to germinate or else is the germination,[77] the continuing 274,1 growth and the generation of like species. For in the case of products of art 'if you were to bury a bed and if the decay were to have the power to put forth a shoot, then it would be wood that grew, not a bed'. This happens because the form is according to custom and 5 convention (i.e. according to the normal practice of craftsmanship as opposed to what is according to nature),[78] and because it is there by convention it comes and goes as something belonging *per accidens*, while the matter persists because it is the essence and nature of the thing, for persistence is the particular property of the essence.[79] But the essence of natural things is according to nature. One could reason thus: in the case of natural things it is the matter and the substrate 10 which persist and generate; such is the essence of natural things; nature is the essence of natural things; therefore matter is the nature in the case of natural things, so that nature is matter. The definitions correspond. Since each form which belongs to the proximate substrate has a substrate of its own, he quite reasonably added 'Similarly, if each of these materials stands in a similar relationship to something else', so that the form changes while the substrate persists; for 15 example, if the bronze and the gold stand in this relationship to water, and bones and wood to earth (and anything else to its substrate), then the substrate would be its nature and essence. For this reason, whatever in each case anyone posited as the primary substrate (whether one or more), that he considered to be the nature and 20 essence of things. Consequently, Antiphon seems to be speaking generally when he says that the substrate is the nature, while whatever each of the others claimed was the primary substrate, that he called also the nature of things. Some said it was one of the elements: for Thales said it was water, Anaximenes said it was air and Heraclitus said it was fire. Others said it was more than one of the elements: for example, Parmenides said it was fire and earth. 25 Others, for example Empedocles, said it was all four of the elements.[80] None claimed it was only earth, although Aristotle, on the analogy of the other elements, made it one of his four.

As for what occurs in the case of the substrate, such as coming to be, destruction and qualitative change, none of these they called nature (for they are not substances), but accidents and affections of substance; those that are transient they called 'dispositions', while the more persistent they called 'states'. They said that the substrate 30 – whichever of the elements it was – was everlasting. For if passing away is a change, occurring in conjunction with the substrate, from something into something else, then the substrate itself could not change from its own nature into something else. Consequently, it could not undergo a destruction; rather, it is everlasting.

If it is destroyed, then all things that have the potential for change are destroyed, i.e. the affections that undergo change in conjunction with the substrate; these they do not call natures. All this leads to 35 the conclusion that matter is nature, that is if they all assume it to 275,1 be everlasting and unchangeable. For it is only right that the nature and essence of all things should persist.

This then, in sum, is the sense of Aristotle's words. Some write 'the formal (*kata rhuthmon*) construction' instead of 'the conventional (*kata nomon*) construction' – which is more comprehensible, since the shape is called form (*rhuthmos*).[81] 5

> **193a28** This then is one sense of the word 'nature' – the primary matter which underlies each of the things which have in themselves the principle of movement and change.

Having said that the word 'nature' is used in one sense as the primary matter which underlies each thing, he adds 'of the things which have in themselves the principle of movement and change'. For even in the products of art, in so far as they are the products of art, there is an 10 underlying matter; for example the bronze underlies the statue, and the stones and wood underlie the house. But these underlie them in so far as they are products of art, and could not be called their nature, since what is being discussed is not something natural but something artificial. Nature must belong to natural things. Consequently, the phrase 'the matter which underlies natural things' could be, in one sense of what has been said, the nature visualised by those making the above assumptions. But in general primary matter – which they 15 want to make nature – does not underlie the products of art.[82] So that even in the case of natural things it is not just any matter that is the nature, but only the primary substrate. This they call 'the ultimate substrate', the substrate of everything else, but having no substrate of its own. For in the case of animals the organic and homoeomerous parts, together with the elements, act as substrate – but none of these can properly be called the nature, because none of them are the 20 primary substrate. Matter is the ultimate substrate even in the case of the products of art, but it does not underlie them as products of art but as natural entities. So it was sensible to say that nature is 'the primary matter which underlies each of the things which have in themselves the principle of movement and change', i.e. every natural entity. For propositions underlie syllogisms as matter, yet they are 25 not their nature, since the syllogism is in no way a natural entity. All natural entities are corporeal. Not even in the case of syllables are the basic elements – the letters – their nature. Consequently, not just any substrate can be called a nature, but only the primary one – and not even that in every case, but only in the case of a natural entity. 30

> **193a30** But in another sense nature is the shape and the form in accordance with the account. For just as that which is according to art and artificial is said to be art, so that which is according to nature and natural is said to be nature. In the former case we would not say that, if anything were only a bed in potentiality and did not yet have the form of the bed, it had anything yet in accordance with art or could even be called art; similarly in the case of natural entities. For what is flesh or bone in potentiality does not yet possess its own nature, until it receives the form according to the account (by defining which we are able to say just what flesh and bone are), nor can it be said to exist by nature. So that in another sense nature could be said to be the shape and form of those things which have within themselves the principle of change – not as something separable other than according to the account.

Having dealt with the hypothesis which claims that nature is the substrate, he puts it on one side and now turns to the one which claims 276,1 that nature is the form. Since nature both shares common ground with art in producing the form and differs from it in so far as nature produces the form in terms of the materials of the art – wood in the case of a bed, bronze in the case of a statue – some people, wanting to make the matter the nature, attempted to prove it on the basis of 5 this difference by saying that the bed, when buried, reveals its nature in the wood and not in the shape, while others, wanting to make the form the nature, pressed the affirmation of their own claim, on the basis of the common ground between nature and art in the producing of the form, as follows: Just as in the case of the products of art, that which is produced by art and is artificial is said to be art (for we say 10 that the shape of the statue is wonderful art), so, in the case of things which exist by and because of nature, that which is according to nature and is natural is said to be nature. For we say that the nature of the wood according to its form is wonderful (if it happens to be so). For just as art stands in relation to the products of art, so nature stands in relation to what exists according to nature and vice versa, as the children of geometricians say.[83] For art is to be found in what exists by art, and nature in what exists by nature.

15 Yet in the case of what exists by art, that which has not yet received the form but still only exists in potentiality cannot yet be said to exist according to art; therefore art does not reside in it. For art resides in the form. For it is not the bronze but the form which is the art of the statue. Therefore even in the case of things which come into being by nature, that which exists only in potentiality is still neither according 20 to nature nor has a nature. For neither flesh in potentiality nor bone in potentiality has the nature of flesh or bone before receiving the

form. Therefore if to have their nature lies in receiving their form, the form would be the nature. One could reason thus: Nature is the very thing whose presence causes what exists by nature to exist by nature; what exists by nature does so by the presence of the form. The form has a twofold character; one according to the shape, the other according to the account. When we define what each is, we render the character that is unique to the shape in terms of the surface configuration, the colour and the size, and the character that is according to the account in terms of the unique formulation of the explicit definition, which corresponds to the definition – as also does the name – and this is what embraces even the shape. Therefore he says that this form – the one according to the account or to such-and-such a shape – is the nature. That is why he wrote 'shape' as well as 'form' when he said 'But in another sense nature is the shape and the form', and again when he said that the form and the shape are not separable. In art it is in another sense that shape and form are the same, because the account of what is artificial is according to the shape. This too is indicative of the fact that the form is the nature. For if the nature of each thing lies in its being, and the being of each thing lies in the form according to the account and the definition (which is why the definitions correspond to what they define), then the nature would be the form. So according to the former arguments matter would be the nature in natural things, while according to the argument just stated it would be the form which is the nature, which is inseparable from the substrate, or rather only separable from it in thought. Separable entities are defined as those which, when separated from what they are said to be separated from, persist in retaining their own nature. But those things for which separation means destruction are not separable, and form in matter[84] is one of these. Natural entities have their being according to just this sort of form, and not according to any separable form.[85]

25 30 35 277,1 5

193b5 That which is a composite of these, for example a man, is not a nature, but exists by nature. 10

Having said that in one sense matter is nature, and that in another sense form is nature, he adds as a consequence in the case of the composite the statement that that which is compounded of matter and form 'is not itself a nature' (for the matter is a nature, as is the form) 'but exists by nature'. For when what is something in potentiality becomes it in actuality, it has a nature and exists by nature. No one says that this is itself a nature. That which has the form which was the nature is said to exist according to and by nature – for example the man compounded of matter and form. For he is not a 15 nature, but exists by nature. Having interpolated this passage about the compound, he adds what else he has to say about the form.

20 **193b6** This rather than matter is nature; for each thing is said to exist when it is in actuality rather than in potentiality. Furthermore a man comes to be from a man, but a bed does not come to be from a bed. That is why they say that it is the wood and not the shape which is its nature, since if it propagates, wood comes to be, not a bed. Therefore if this is art, the form is nature, since man is propagated by man. [This passage is not printed as a *lemma* by Diels]

He adds this comment about form since he wants both matter and form, but more so form, to be nature. He adds the reasons why he wants form rather than matter to be nature, demonstrating it in several ways. But Porphyry[86] thinks that he is referring to the 25 compound when he says 'This rather than matter is nature'; even if this is not nature in the proper sense but is only by nature, nevertheless it is nature to a greater degree than matter is, since it possesses within itself the form, which is nature[87] to a greater degree. The first proof[88] goes like this: Nature is the reason why each natural thing is what it is said to be; the reason why it is what it is said to be is the reason why it is in perfect realisation (*entelekheia*) and not merely in 30 potentiality; the reason why it is in actuality what it is said to be is the form; therefore its nature is the form. Alexander summed the argument up as follows: Each existing thing is what it is when it is in perfect realisation; so what exists by nature so exists when it is in perfect realisation; anything is in perfect realisation when it pos- 278,1 sesses the form, and so things that exist by nature so exist when they possess the form; but the very thing whose presence causes what exists in nature to so exist is nature, and it is by the presence of the form that what exists in nature does so exist; therefore the form is nature. By adding the word 'rather' Aristotle showed that even what is in potentiality can be spoken of <as existing by nature>, even if not in the same way as what is in perfect realisation; consequently, 5 matter is nature, even if to a lesser degree than form is. The term 'perfect realisation' is said to be Aristotle's own. It signifies the form which is in actuality in as much as it is in this respect that we refer to the reception of the one perfection, or to the reception of being one and perfect, or to the continuous possession of the perfect, i.e. the state according to perfection.[89]

10 Aristotle further mentions those who claim that the matter, not the form, is nature; if they make this claim because when the wood is buried any growth that occurs from the wood is wood, not a bed, then nature would be the form (since a man is propagated by a man,

and a man is a man because of the form). Even if a bed is not propagated from a bed, a man is propagated from a man like wood 15 from wood, and in general products of art are not propagated from other products of art, but natural things are. Since this is the case, then anyone making a general rule about natural things must say that the form is the nature; but he should not disavow it in the case of the products of art. For art does not make forms that are productive of their like, as nature does. For even if the wood is the matter of the bed, it is still a natural form and it is in respect of this that it too has the power of propagating its like. The words 'A man comes to be from 20 a man' were proclaimed like some final conclusion, but only served to cloud the argument. The gist of the argument is like this: A man comes to be from a man as a natural form from a natural form, not as an artificial form from an artificial form. Looking to this latter they claim that the form is not nature, but they ought rather to look to natural entities that propagate things like their own forms; since this 25 is peculiar to nature, they should say that the form is nature. For if matter, by persisting as matter in the propagation of wood from wood, seems to them to be nature, then let the form too seem to them to be nature by its persisting. It is worth considering whether, when a man is propagated by a man, the form is propagated by the form, and not the compound by the compound.[90] Now this is why he said that. But 30 if this is the case, then the shape is nature, since the shape too comes to be in the compound.

> **193b12** Again, the word 'nature' used in the sense of 'growth' is a process towards nature. It is not like medicine, which is spoken of as process towards health, not towards techniques of medicine. For medicine must start from such techniques, not aim at them. But growth does not stand in this sort of a relationship to nature; what is growing, in so far as it is growing, goes from one thing into another. What then is it growing into? Not what it started growing from, but what it is growing into. The form therefore is its nature.

This third proof that he offers to show that nature is the form is very 35 skilfully presented. 'Nature' used in the sense of 'growth[91] and coming-to-be' is a process towards nature, starting from the thing that is 279,1 coming-to-be, and finding completion in its nature. Just as in the case of the products of art anything that is being made[92] is said to be being made into what lies at the end of the process, not what lay at the beginning (when a carpenter is making a bench out of wood, it is the bench, not the wood, that is said to be being made by him), and similarly in the case of natural things (when water changes into 5 vapour, we say that it is the vapour, not the water, that is being

made); just so, then, the thing which is growing is said to be growing when it is proceeding towards its nature, not from its nature. It is proceeding towards its form; the form therefore is its nature. The summary of the argument is as follows: Nature is that towards which
10 anything growing and increasing is proceeding; anything growing and increasing is proceeding towards its form, not towards its matter; therefore its form is its nature.

But sometimes something coming-to-be gains its name not from the end of the process but from what endows it with the form towards which the process is aiming – for example medicine. The term 'process', by its very character, can be applied just as well to nature
15 as increase; but medicine is not a process towards medical techniques in the way that nature is a process towards nature – rather it is a process towards health, which is produced by medical techniques, for it is the patient who is being healed. That is why Aristotle made the distinction, pointing out that the way nature is related to nature as process to goal is not the same as the way medicine is related to health. He needs to make this distinction in terminology because in
20 most cases the cause and the effect are called by the same name: heat causes heat, and the process is called heating. In some instances the names differ: medical techniques cause health; perhaps if we were to call it healing instead of medicine it would then be similar in character. For just as nature as coming-to-be is a process towards nature as a goal (for nature is the goal achieved by whatever is growing), so healing is a process towards health. The matter is that which is
25 growing; what it is growing into is the form and the goal, not the starting point. If, then, that which is growing receives the form, it can in this case again be said that that which is growing receives a nature. The form, then, is a nature according to this third argument presented, which is convincing only in part; for in it nature is a form, but not something natural. But Aristotle, having said that what it is
30 growing into is the goal, not the starting point, failed to add that the goal is a form and a nature (because it is obvious), but he did add the conclusion 'The form therefore is its nature'. And having said 'goes from one thing into another' he added 'in so far as it is growing' to demonstrate that 'goes from one thing into another' is the same as 'grows from one thing into another'. For the goal that it is going towards is what it is growing into, not what it is growing out of, either
35 as the substrate or as the opposite, for both of these are starting points.[93]

280,1 **193b18** 'Form and nature' can be used in two different senses. For privation is a sort of form. We must consider later[94] whether or not privation and an opposite are involved in simple coming-to-be.

He has said that nature as a process reaches its completion in nature and form. Since process is twofold – the one from privation to form, 5 the other from form to privation – such a nature according to process would also be twofold, as would be the goals – form and privation. If, then, the goal is form and nature, how could privation, which is the opposite of form, be a goal? Further he posited, as the elements of things in process, matter and the opposites.[95] So since he has now posited nature as one of the opposites,[96] there was an obvious problem as to how both of these were causes, but only one of them nature. That 10 is why he granted to privation also a share in nature, to the same extent that he granted it a share in form, in the words 'For privation is a sort of form'. It is a sort of form, either because it is distinguished from the substrate as something opposed (since it too resides in the substrate as the form does, and it too becomes a sort of shape and form of the substrate, for the accidents are forms of a sort in their subjects), or because privation is not absence pure and simple, but is 15 in something that is by nature <such as to admit the corresponding form>. So the suitability for this form is a faint image of it. That is why it gives a certain disposition to the substrate – which is the characteristic of the form. Or else not all privation could be said to be 'a sort of form', but only that which is viewed as the inferior of a pair of opposites; for this form is subsumed under privation, as he demonstrates in *On Coming-to-be and Passing Away*, and privation taken 20 in this sense is form.[97] In fact he says there that one of the opposites – whichever is taken – is the privation of the other. If, then, privation is a sort of form and a nature, being something other than the substrate and the matter, even if it has its being in the matter, even things that exist in privation could be said to exist by nature – but not according to nature, since the aim and the goal of the substrate 25 exists according to nature, and its aim is the form. Therefore because of privation what exists by nature covers a wider field than what exists according to nature.

But how could the process towards privation be called a nature, since it is not a growth? For it is a passing away, not a coming-to-be. Well, in so far as privation is a sort of form, both it and the process leading to it are 'nature'. But if privation is a sort of form, it is clear 30 that form comes from form, and we can no longer say that the highest type of opposition is that of form and lack, as was claimed in book 1.[98] But it is quite clear, and was shown by analysis in book 1 that even if form comes from form, it does so not as from form but as from privation, and from what is not such, but is by nature such as to be such. But if 'privation is a sort of form' it will be opposed to form not 35

just as privation but also as an opposite. For the opposition of forms that cannot coexist depends on their being opposites, not on being a
281,1 state and its privation.[99] If this is the case, all coming-to-be is not just from what is not such, but from opposite forms. And if this is the case, there will either be no coming-to-be of the substance, and every substance will be ungenerated and imperishable (although among natural things that which is, according to him, properly substance, <i.e.> individual substance, does come-to-be and pass away, and that
5 which is properly called a coming-to-be according to the substance is a change), or else, if a substance does come-to-be, then it must come-to-be from some opposite. But Aristotle thinks that a substance has no opposite. For the present he postpones this question which causes problems on either side, whether or not, since privation has been called a sort of form, there will be any privation involved in
10 simple substance, if privation is an opposite form. For if not, since all coming-to-be is from an opposite, the coming-to-be of a substance – which, according to him, is the only proper coming-to-be – will be an impossibility. But if there is a coming-to-be of a substance, there is some opposite to substance, which is what he denies in the *Categories*.[100]

Elsewhere he solves these problems by saying that there is a substance whose coming-to-be seems to be the compound made up of
15 matter and form, but that this does not come-to-be simply but in respect of something. In respect of the substrate it is ungenerated (for matter is neither generated nor destroyed, as he proved at the end of book 1), but it is generated in respect of the compound, i.e. the substance in respect of its form which resides in the underlying matter. Just as a substance has generation, so there is something
20 opposite to it. But it has generation not in respect of the substrate, but of the form residing in the substrate. In this respect, then, it does have something opposite, not in so far as it is a substrate, but in so far as it resides in a substrate. For fire does not have an opposite in so far as it is a compound and this sort of a substance, but it does in so far as it is a form. It has opposites in respect of its heat, its dryness and its tendency to rise upwards, and its generation is in respect of
25 these. So when he says in the *Categories* that substance has no opposite, we should understand him to be talking of the compound. For in that book he is talking about substance in this sense alone, together with its genera and species. That is why there is no opposite to substance simply as substance, nor is substance as substance generated. For it would be generated from non-substance and what,
30 in the unqualified sense, does not exist. But in so far as a substance is of such-and-such a kind and there are different substances, there
282,1 is opposition within them. In fact it is in this respect that there is

coming-to-be; differences in substances are according to their species, so that it is in these that the difference lies.

'Either', says Alexander, 'there is nothing opposite to substance in respect of the form – for the privation from which the form comes-to-be is not properly an opposite – or else the privation too would be an opposite if one of the opposites is a privation.' This is what Alexander says. But perhaps coming-to-be is not from privation of this sort in the strict sense, the one said in respect of the inferior opposite. For even if coming-to-be is from the opposite form, it is not in so far as it is a form, but in so far as the privation of the opposite coexists with it, being an absence in what is by nature <such as to admit the form>; for the same thing is by nature such in relation to each of the two opposites. For just as a common genus precedes the opposites, so a common substrate receives them. Perhaps, then, in another sense there is no opposite to a substance in respect of the form, in so far as the opposites are considered in the light of the contrast of the differences between them, while the form is not a difference but a conjunction of differences within the genus. The opposites are qualities, but the form is a substance. If, then, there is no opposite to the form, how can the form come-to-be? Therefore either it is not generated but together with the matter is ungenerated, and the compound is ungenerated, and everything is ungenerated, or else not all things that come-to-be come-to-be from opposites. Perhaps, then, not even the form *per se* comes-to-be or passes away. For fire is ungenerated not only in respect of the compound (as was stated earlier) and the matter, but also in respect of the form. For it is generated in relation to the differences which constitute it, in which opposition too is observed. For of its heat, its dryness and its tendency to rise, none is the form of fire *per se*, but these are its differentiae. All of them combine to produce fire. Fire therefore comes-to-be and passes away according to the generation of these from their opposites and their destruction back to the opposite. In the same way in the case of the other forms too the generation and destruction of each does not belong to the form *per se*, but goes according to the differentiae of the form. When they are destroyed by each other, the form is destroyed together with them, just as it supervenes when they combine.[101]

5 10 15 20 25

<CHAPTER 2>

193b22 Now that we have determined in how many ways nature is spoken of ... 30

Aristotle himself has made it quite clear, by summing up his argument with the words 'Now that we have determined in how many ways nature is spoken of', that the whole thrust of his argument has

been directed at just that – to distinguish the different meanings of the term 'nature' – since thinkers have understood the term differ-
35 ently according to the different meanings. The phrase 'in how many
283,1 ways' is applied to things different in their substrates. But since he has given a clear exposition of the other meanings while keeping the principal one concealed, it would be a good idea to review them all briefly, saying just this by way of introduction, that since natural body comprehends matter, form and the compound, and is generated and consequently embraces both the change which results in coming-
5 to-be and above all the cause of change (for where there is change there is in all cases a source of change), nature can be spoken of in five ways.

1. According to one meaning it is the matter of each thing, that which belongs primarily as something formless to every natural entity, just as in the case of the products of art, in the statue it is the bronze, in the ship it is the wood, in every natural body it is the
10 primary matter (describing it from the bottom upwards) or the ultimate substrate (as those starting their analysis from the top call it).[102] This seems to be nature because the nature of each thing must be seen to remain the same throughout all its various changes. Thus the nature of a man is what is clearly seen to be the same in all respects when he is awake, asleep, moving, at rest, and undergoing
15 any other change. So in this way, in all the various changes of any natural body, that which remains the same would be its nature. And it is the matter which remains the same. Antiphon even tried to prove the persistence of the matter also from the sprouting of like material, although the sprouting demonstrates that form comes from form rather than matter from matter. For man is propagated by man, and
20 wood by wood; the wood too is the form, even if it has the description of matter in relation to the bed.[103]

2. According to the second meaning, nature is said to be the form which is to do with the matter. For just as the statue is not called a statue according to the terminology of art until it has received the form according to the art, just so the matter is not called by the name of any natural entity until it has received the form. For the matter is
25 only potentially the thing of which it is the matter, as for example the seed is only the animal in potentiality, and each thing receives its specific designation only according to what it is in actuality. That is the form. So nature quite reasonably seems to be form rather than matter.

3. According to the third meaning, nature is said to be the compound of matter and form, for example the man. For just as the word 'substance' is used in three ways – matter, form and compound – so
30 'nature' could be spoken of in three ways. But Aristotle says of the compound: 'it is not a nature, but exists by nature'. For if each of the

two components is itself a nature, and if the compound which exists because of them is something else other than these two, then it would not strictly be a nature, but only exist by nature. But if the words 'This *rather than* matter is nature' are said of the compound, as Porphyry understood them, it is clear that the compound is not a 35 nature in the strict sense (for not even any of the simple bodies is a nature in the strict sense) but is a nature more so than matter is because it possesses within itself the form, which is nature more so than matter is. Moreover according to Antiphon's distinction, since a 284,1 man is propagated by a man as compound by compound, the compound too would be a nature. These three meanings of nature comprise the compound and its components.

4. According to the fourth meaning, nature is said to be, as it were, 5 the growth, coming-to-be and change by which the growing thing is made to grow by that which produces its growth. For just as in the case of a cloak it is the weaver who weaves, the cloak which is woven and (as a third thing) the act of weaving which is a change stemming from the agent and directed towards the artefact, just so in the case of a natural entity there is what is growing, what makes it grow and, between the two, such a nature as is the change stemming from the 10 active nature – as the process of healing stems from the art of medicine.

5. According to the fifth meaning, the most important one, nature is the cause of change in natural bodies. Just like art (which is the producer of artefacts) and its motive force, nature in this sense starts from the material nature and ends at the formal nature, producing 15 the compound nature. The productive nature has this much in common with art, but differs from it in the fact that the art, which is external, starts from the considerations proper to it and ends at some completion beyond itself. For in this way the end of medical skill is health, but the nature which is inherent in what is growing works through the, as it were, outgrowth towards the nature of the perfected 20 entity, and ends as a nature reaching a nature through a nature. So even if art preserves its similarity to nature in this respect by working through artistic change and ends up with the artefact which is of like form, even so nature differs from art in that the actualisation of the nature is inherent and internal.

These, then, are the various meanings of nature. The word itself 25 is more strictly suitable to nature in the sense of change and growth (on the analogy of weaving, healing and change in general). But the common connotation of the word fits better with nature in the principal sense, i.e. the nature which is productive of natural things. However, the definition will suit all the meanings if taken of each in the appropriate way. For nature in the strict sense is a principle and 30 cause of movement and its cessation, while nature as change is something instrumental; for it is by means of nature in this sense that the productive nature brings to completion change and its cessation in natural things, just as a doctor brings about health through the practice of medicine. Matter and form are sources in an elemental sense of the actualisation brought about according to 35 nature. Eudemus says that even these admit being described as 285,1 nature; for the matter and the end in view seem to be sources of change,[104] since we say that the underlying matter is responsible for the fact that lead drops downwards; it is carried downwards because it is made of this sort of matter; hence it has a principle of change within itself and *per se*, for it is lead *qua* lead. But the form is a principle also in the sense of an end in view, since it is to this that 5 nature looks in doing all its work. But how could the compound be a principle and a cause when it comprises only the finished product? The answer is that even this could be a principle in the sense of an end in view. For whether the compound is a form in matter or a product of matter and form, nature is productive of this compound and not of the form as existing *per se*. But perhaps the compound is a principle of change and its cessation also as a productive cause, 10 since actualisations and their cessation are produced in compounds as changes. For the substance of fire would be a principle and cause of its ability to heat and its tendency to rise upwards. But enough of that.

It seems worthwhile asking why Aristotle, in his listing of the different meanings of nature, failed to give the most important one, 15 that of nature as productive of natural things. He said that nature is the name given to the matter and the form, and then (since this is granted) to the compound and to the outgrowth (so to speak) and the change leading to the form; but he made no mention of it as the productive cause. In reply I think it must be said that at the very outset of his discussion about nature he presented nature in the strict sense as the productive cause and defined it as such.[105] That is why, 20 when he was explaining the term '*per accidens*', he took as an example the doctor treating himself, as though seeking that which is a *per se* productive cause, that which is analogous to the person making a house and other artefacts. So having in that passage presented nature in the strict sense, he here offers the other meanings of the description of nature. And perhaps he did not even in this passage pass over nature in the strict sense, but revealed it when he said that 25 healing was a process starting from medical knowledge and aiming not at that knowledge but at health, and that nature as change started from a nature and moved towards a nature. The nature that is analogous to medical knowledge is the productive nature and not any of the other kinds of nature that correspond to the four meanings.

He will, in my opinion, have something to say about productive nature a little further on when he distinguishes the causes.

But since the discussion of nature seems to have come to a conclu- 30
sion, it would be a good idea to resume the argument and to ask what, according to Aristotle, nature is and what power it possesses among existing things. He was right to look for it in the difference between what exists by nature and what does not. For it was thus that both Aristotle himself and Plato discovered the soul, from the difference between what is ensouled and what is soul-less, when he said in the 35
Phaedrus:[106] 'All body which is moved from outside itself is soul-less, 286,1
while what is moved from within itself is ensouled, this being the nature of soul.' And in the *Laws*[107] he says that that which is moved from within itself we say to be alive. It is in this respect that the ensouled differs from the soul-less, which Aristotle too in the second book of *On the Soul*[108] says in almost the same words: 'Making a new 5
start for our enquiry we say that the ensouled is distinguished from the soul-less by being alive.' The phrase 'what does not exist by nature' has two senses: in one sense it comprises what is above nature, such as the immaterial, the bodiless, that which is apart from body lodged in pure forms, and in the other sense it comprises the lower order of natural things, things produced by art such as a bed or a cloak and similar enmattered bodily things. What is common to 10
things above and to things below nature is the fact that they are not moved or changed in themselves,[109] although the one kind <has these properties> for the better, the other for the worse. Natural things lie between these two orders. Because they lie below the whole of immaterial and bodiless being they are themselves enmattered and bodily; because they are not produced by any art, but (as it were) grow and increase by themselves (their creative cause being imperceptible 15
to the senses), they are called natural; because they differ from the upper and lower order they have within themselves a principle of movement and change. So it was quite reasonable for Aristotle to present this as a characteristic of nature and its essence, by saying that it is a principle of movement and change and of the cessation which terminates such change.

But the soul too is a principle of movement and change in ensouled 20
bodies according to both Plato and Aristotle himself. What, then, is the distinction? For it has been stated previously and should now be reiterated that even the lowest part of the soul, called the vegetative,[110] is something other than nature even according to Aristotle[111] – even if they[112] often call the vegetative part of the soul nature on the grounds that it is close to nature. All soul, even the lowest kind, 25
is said by him to be 'the ultimate actuality of the natural body possessing organs', so that the vegetative part of the soul belongs to the body which has a nature, and so is something clearly different

from its nature. But it is not only bodies with organs that have a nature, but also homoeomerous substances and the four elements. Furthermore, we give the name 'ensouled' to things that have within themselves the cause of growth, increase and the propagation of their
30 kind, while we designate as 'natural' also things which are not like this, such as rocks, other minerals, lifeless bodies and simple bodies. Furthermore, all body has a nature (including the materials of artefacts like the material of the statue) and is natural just like the wood of the bed. But not all body is ensouled. Therefore nature would not be soul. It is clear that nature is inferior to the vegetative soul,
35 since such soul supervenes on natural body as form on matter. How then did Aristotle present the difference between it and soul? I think
287,1 that the phrase 'in which it resides' is sufficient in this respect, and the following clearer phrase 'in a substrate' refers to nature. For all soul, because it properly has the power of moving, is set apart from what is moved. If this does not satisfy those who consider that the vegetative and irrational parts of the soul reside in bodies as sub-
5 strates, then at any rate the following will satisfy them, since it is most decisive both for the understanding of the natural substance and for the distinction between it and the soul. Aristotle does not say that nature is a source of change for bodies in the same sense that both he and Plato say that the soul is. For according to both the soul is what moves bodies, but nature is not a source of movement in
10 respect of moving but of being moved, and of cessation of movement not in respect of stopping but being stopped. That is why natural things are not said to be moved by themselves. For if, says Aristotle,[113] they could move themselves, then they could also stop themselves. Nature seems to be a sort of propensity for being moved and regu-
15 lated, as it were growing upwards from below and because of its own fitness inviting regulative causes. For if it were a principle of movement in the sense of causing movement, it would not in this respect differ from the soul and the primary moving cause. But since bodies are far removed from the indivisible and unextended essence and the life that subsists in absolute being, and are lifeless and spiritless in
20 themselves, too chilled for any kind of life, they have within themselves the most distant sort of form of life which relates to potentiality and propensity, that which we call nature, because of which even lifeless bodies can be moved and changed, and what is more grow and act on the passive in each other. For their activities are not pure, but involve being acted on. That is why all natural bodies can only move other things if they themselves are moved. Yet what is strictly
25 unmoved itself moves, as Aristotle says.[114]

That Aristotle defines nature as the principle of change in the sense of being changed, not causing change, he makes clear when he says in this work[115] 'nature is a principle and cause of being changed and its cessation' and that the nature resides in the substrate. But that which resides in the substrate would not in the proper sense be a principle that moves the substrate. However, in the last book of this treatise,[116] when discussing the four elements, he says: 'It is clear that none of these things moves itself. But they do have a principle of movement, not in the sense of acting or causing movement, but of being moved.' That is why he asks what it is that moves the elements, since they are not moved by themselves. For he wants this to be particular to animals that have a soul, which he defines as a source of movement. Before that passage he says:[117] 'Similarly that which can by nature be changed is that which is potentially of a certain quality, quantity or position when it has within itself such a source' – clearly referring to its nature. In book 2 of *On the Heavens*[118] he wrote the following: 'In none of the soul-less entities can we see the origin of the source of movement. For some are not moved at all, others are moved but not similarly from every side – for example, fire which is only moved upwards, and earth which is moved towards the centre.' If, then, the four elements are natural entities and do not have within themselves the origin of the source of movement, i.e. the moving cause, it is clear that it is not in this sense that nature as a cause of movement is said to be a source of movement, but as a source of being moved. If, then, nature is something like potentiality and the propensity to be moved, why do we so often say that nature is active? In this very book Aristotle himself says[119] that it is analogous to art, and towards the end of the book he shows that nature acts for the sake of something. Summing up the argument he says:[120] 'It is obvious, therefore, that nature is a cause and a cause in the sense of being for the sake of something.' In book 1 of *On the Heavens*[121] he is clearly matching nature's work with that of God, and says: 'God and nature do nothing to no purpose.'

30

35
288,1

5

10

15

I think we should say in answer to this too that everything that comes-to-be does so from a substrate which is potentially that which it is going to become, and which actually is that by the agency of whatever produces the change. Both are necessary for the end result. For this reason, even if nature is a propensity in the substrate, it is said to act because it contributes to the end result; and when he says that it acts for the sake of something, he means that the coming-to-be of natural things looks towards a definite goal, and that they do not come-to-be through luck or chance, but because they are constituted by nature to become what they become. At any rate he says in this book:[122] 'Where there is a goal, the other[123] and successive stages are completed for its sake. Therefore as each thing is done, so it is constituted by nature, and as it is constituted by nature, so it is done, if nothing prevents it.' So you can see that what he means by 'that which is constituted by nature' is the natural entity. So even if he

20

25

says 'God and nature do nothing to no purpose', he means it in the sense that nature provides from below the propensity which looks towards a goal which is the good, while God sheds his light from above 30 in the form of actualisation. In this way Aristotle revealed, from the difference between what exists by nature and what does not, the substantial being of nature.

Earlier thinkers too clearly had some such conception of nature viewed in terms of the propensity of each thing for change, according 35 to which natural entities are characterised. But since all natural 289,1 entities have matter and form, some ascribed such a potentiality to the matter, claiming that this was a nature by which natural entities are constituted to be changed, and seeing that they were most changed in their matter, as for example a bed is changed in terms of the wood. Others, claiming that it was nature in terms of which 5 natural entities have their being, since the form is the mark of each thing by which each thing subsists and is said to be just what it is, for this reason said that the form was the nature. Because of this conception of nature, the one that goes by the character of each thing, we use its designation in the case of everything, and do not eschew talking of the nature of soul and mind, and even of God. But Aristotle 10 did not think it right either to call matter *per se* either nature (for matter *per se* is an impotent substrate), or to call the form <nature> (for this is natural but not nature), but he designated as nature the propensity of matter for appropriate movement and change, when it changes from one form to another. For the loss and reception of the form happens to matter according to its natural propensity. The form 15 comes-to-be according to its own nature from its opposite, and once it has come-to-be it is preserved and changed by both acting and being acted on, or rather by being actualised passively. So matter and form are both natural, but neither is nature; nor, similarly, is the compound. Form rather than matter would be nature because of its character and power. And the compound rather than matter would 20 be nature because of the form, since it becomes an entirely natural individual thing when it receives the form. For matter *per se* is lacking in definition, while nature, being a propensity for the existence of the form, in a way pre-exists the form by being present in the matter in potentiality, as you would expect; and it gives within itself a prior indication of the form, being its nature and its growth, as it were, and 25 its sprouting from the matter. So those who say that nature is the lowest level of life are quite right. For just as the bubbling over, as it were, from primary Being into the separating out of the Hypostasis of Form and the movement outwards from Being to Actuality is the primary potentiality and the primary life which subsists according to the first movement of Being, just so the growth of the enmattered 30 form from matter and the movement towards that form viewed according to the potentiality of the form is the lowest potentiality and the lowest life.[124] For this reason Being is higher, above life, and matter comes lower after nature, because the higher causes far outstrip the inferior. Nature, being the life of the form, is not only its growth, but also its continuation once it has come-to-be, and its rising up to act and be acted upon according to its natural constitution. 35

> **193b23** We must next consider in what way the mathematician differs from the natural scientist. For natural bodies too have planes, solids, lengths and points, all of which are the concerns of the mathematician. A further question is whether astronomy is or is not a part of natural science. It would be absurd if it were the concern of the natural scientist to know just what the sun and moon are, but not to enquire into any of those properties which are *per se* consequences <of the definitions>, especially because people who discuss nature obviously discuss the shapes of the moon and sun as well, and also whether or not the earth and the cosmos are spherical. 290,1
>
> The mathematician too is concerned with these features, but he does not consider each as the boundary of natural body; nor does he consider their properties as properties of them as physical bodies. That is why he separates them; for he can mentally separate them from the process of change, and it makes no difference and no errors are caused by such separation.

He quite justifiably wants to show the difference between the natural scientist and the mathematician, since they appear to concern themselves with the same subjects. The natural scientist is concerned with 5 natural bodies, which 'too have planes, solids, lengths and points, all of which are the concerns of the mathematician'. If these features belong to the natural world, then the mathematician would be a natural scientist, and mathematics would be a branch of natural science, since it too is concerned with the natural world. He presents the difference between natural science and astronomy in a similar way, since it too seems to share common ground, particularly because 10 the astronomer, like the natural scientist, is concerned with moving bodies, and not with motionless bodies, like the geometrician. That is why the natural scientist and the astronomer appear to deal with the same subject matter. For even if the natural scientist discusses the substance of the heavenly bodies, while the mathematician discusses their properties, such as movement, shape, size and the 15 distances that separate them from each other and from the earth, even so the natural scientist discusses these features as well. For it would be absurd for the person who thinks that he has conducted an

enquiry about the nature of reality not to know about these features. Furthermore, those who concerned themselves with a study of nature clearly spoke about 'the shapes of the moon and sun', and showed that 'the earth and the entire cosmos were spherical'. Moreover, Aristotle
20 himself in *On the Heavens* and Plato in the *Timaeus* both used the arguments of natural science to demonstrate truths about these matters. If, then, the natural scientist conducts his enquiries about both the substance of the heavenly bodies and their properties, and the astronomer only about their properties, astronomy too would seem to be a branch of natural science, and in this way both mathe-
25 matics and astronomy would be very close to natural science, which is why he needed to present the difference between them.

The mathematician differs from the natural scientist in the first instance in that the natural scientist talks not only about the properties of natural bodies but also about their matter, while the mathe-
30 matician is in no way concerned with the matter. Secondly, they do not talk about the properties which are the common subject of their discussions in the same way; the natural scientist talks about planes, lines and points as boundaries of the natural body subject to change, while the mathematician does not include in his considerations the inherent tendency to change nor does he take the above features into
291,1 account as boundaries of a natural body. For not even is the solid assumed by him to be natural <body>, but only something with three dimensions as if such things existed *per se*; for the mathematician concerns himself with the features that can be mentally separated. That is why he separates them, and there is no consequent error. For having mentally separated them from natural bodies and all change he examines in this way the consequential properties of the things
5 that are thus assumed. Therefore if he talks about them not as natural entities nor about them as properties belonging to natural entities, then the mathematician would not be a natural scientist, nor would mathematics be a branch of natural science. Nor would astronomy; for it too deals with the properties of natural entities not *qua* properties of natural entities, but as the sort of things that are shaped
10 and moving bodies. So that the astronomer too, even if he talks about properties that belong *per se* to natural bodies, does not enquire into them *qua* belonging to natural entities, nor does he seek to prove that such shapes, sizes and movements belong to such a nature; in just the same way, with regard to the fact that the body of the heavens is spherical, the natural scientist shows that this is the case from the fact that of all the solid shapes sphericity alone is primary, simple,
15 perfect and one in form (for rectilinear bodies are compounds of many factors and are secondary), and for this reason that it belongs to the first of bodies,[125] as Aristotle proves; but the astronomer proves it from the fact that among solid bodies the sphere is more capacious

than any other body with an equal surface area. In this way, then, Aristotle demonstrated succinctly the difference between natural science on the one hand, and mathematics and astronomy on the other. 20

Alexander quotes in a scholarly manner a passage from Geminus' abridgement of his commentary on Posidonius'[126] *Meteorologica*, which shows the influence of Aristotle: 'It is the concern of natural science to enquire into the substance of the heavens and the heavenly bodies, their powers and the nature of their coming-to-be and passing away; by Zeus, it can reveal the truth about their size, shape and 25 positioning. Astronomy does not attempt to pronounce on any of these questions, but reveals the ordered nature of the phenomena in the heavens, showing that the heavens are indeed an ordered cosmos, and it also discusses the shapes, sizes and relative distances of the earth, the sun and the moon, as well as eclipses, the conjunctions of heavenly bodies, and qualities and quantities inherent in their paths. 30 Since astronomy touches on the study of the quantity, magnitude and quality of their shapes, it understandably has recourse to arithmetic and geometry in this respect. And about these questions, which are 292,1 the only ones it promised to give an account of, it has the power to reach results through the use of arithmetic and geometry. The astronomer and the natural scientist will accordingly on many occasions set out to achieve the same objective in broad outline, for example, that the sun is a sizeable body, that the earth is spherical; but they do not use the same methodology. For the natural scientist 5 will prove each of his points from the substance of the heavenly bodies, either from their powers, or from the fact that they are better as they are, or from their coming-to-be and change, while the astronomer argues from the properties of their shapes and sizes, or from quantity of movement and the time that corresponds to it. The former, with an eye to productive power, often touches on causes. But the 10 latter, when he is constructing proofs from what impinges on a heavenly body from outside itself, does not pay any great attention to causes, as for example when he is presenting the earth or the heavenly bodies as spherical; sometimes he does not even attempt to find the cause, for example when he is discussing the eclipse; on other occasions he makes his findings on an assumption, demonstrating certain orbits[127] by whose presence appearances will be saved. For 15 example, if, in reply to the question: "Why do the sun, moon and planets clearly move in an irregular manner?" we assume that their orbits are not centred on the earth or that the heavenly bodies have their poles on an epicycle, their apparent irregularity will be preserved and it will be necessary to enquire closely in what ways it is possible for these appearances to be produced so that the treatment of the planets is squared with the accepted method of causal expla-

20 nation. That is why a certain Heraclides of Pontus[128] came forward to say that the appearance of irregularity in the sun's path can be saved if the earth moves in some way, and the sun stays still in some way. In general it is not the concern of the astronomer to know what by nature is at rest and what by nature is in motion; he must rather make assumptions about what stays at rest and what moves, and 25 consider with which assumptions the appearances in the heavens are consistent. He must get his basic principles from the natural scientist, namely that the dance of the heavenly bodies is simple, regular and ordered; from these principles he will be able to show that the movement of all the heavenly bodies is circular, both those that revolve in parallel courses and those that wind along oblique circles.' 30 In this way Geminus, or Posidonius as presented by Geminus, displays the difference between natural science and astronomy, showing the influence of Aristotle.

In this passage Aristotle says that it would be absurd for the natural scientist to know nothing of the essential properties of natu- 293,1 ral bodies; this is a fair claim, since the person who knows about anything properly is bound to know its particular characteristics and its proper differentiae. For if he does not know these, he will not even know whether it is different from things of its own type. If he does not know this, he will suppose that all things of the same type are the same as each other, so that if this is the case, <he will identify> 5 even the opposites. If, then, the shapes, sizes and such movements of the heavenly bodies are particular to divine body, the natural scientist is bound to know them as well. With regard to verbal usage, it is worth noting that here he seems to call a geometer a mathematician; for he is the one concerned with planes, solids, lengths and points, although mathematics proper also includes arithmetic, astronomy (or the geometry of spheres) and harmony (or musical theory). Since 10 'astromancy' had apparently not yet arrived in Greece, the ancients applied the name 'astrology' to what is now termed 'astronomy'; more recently people have made a distinction in the terminology and have been calling the study that looks to the movements of the heavenly bodies 'astronomy', and they have given that which deals with the 15 results of those movements – 'astrology' – its own particular name.

> **193b35** Those too who proclaim the Forms do the same thing without realising it. For they separate natural entities, which are less separable than mathematical entities. This would become clear if you were to attempt to define the entities themselves and their properties. For the odd, the even, the straight, the curved, as well as number, line and shape will be found to be independent of change, but flesh, bone and man will not be –

> these are spoken of as a snub nose is, and not as what is curved. The point is also demonstrated by the more physical branches of mathematics, such as optics, harmonics and astronomy. They are, in a way, related inversely to geometry; for geometry deals with natural lines, but not *qua* natural, while optics deals with mathematical lines, but *qua* natural, not *qua* mathematical.

He has said that mathematicians mentally separate things that are not separable in reality, and construct their arguments about them in this light. Because they abstract the three dimensions from natural 20 body, and of these three dimensions they take two as a plane figure by itself, and of these two dimensions abstract one (when they are dealing with the line), he now says that those who posit the Forms[129] are doing the same thing, although they do not realise it; in fact they are doing something even more absurd. For mathematicians separate lines, planes and suchlike from natural bodies; even if these things are inseparable in reality, nevertheless, when they are mentally 25 separated, no impossible consequences follow. But those <who posit the Forms> attempt to separate the natural forms such as man, flesh and suchlike, which are less susceptible to separation than mathematical entities. He seems to use the word 'less' because of philosophic caution, although he wants these things to be in no way separable, even mentally. He offers a general rule to determine what can and 30 what cannot be mentally separated. When, in defining what we are separating, we do not include in the definition the entity from which we are separating it, and do not carry it along in our conception, but instead define and conceive it as something *per se*, it is then that we say such a thing is separable in definition and thought (for example, when defining the mathematical body we talk about that which has three dimensions without in any way carrying along the matter or 35 the movement of the natural body; in defining the plane surface we talk about that which has nothing more than length and breadth; and it is the same in the case of numbers). But when the original entities 294,1 appear inevitably as part and parcel of the definition which we seek to give, together with the properties which we are separating, which cannot even be thought of without them, then we say that such entities cannot be separated even in concept and thought. Such entities are flesh, bone and man. For not only is each of these natural, 5 enmattered and a composite, but each is also thought of as a compound. He showed the difference by taking the example of shape, which can on occasions be thought of by itself, as when it is spoken of by itself as convex or rather as concave, but which on other occasions cannot, as in the case of what is snubnosed. In fact snubnosedness is concavity, but not just any concavity, but the concavity of a nose. Therefore snubnosedness cannot be thought of apart from 10 a nose; in giving the definition we say that snubnosedness is concavity of a nose. If, then, when defining man as a mortal animal possessing powers of reason we speak of what is called the model of man according to the form, either something other <than man> is called 15 'man' (for not even his model would share anything by way of substance with the naturally constituted man), or else, if the ideal man is a mortal animal possessing powers of reason, and if it is to him that the man down here on earth owes his being of like nature, through his similarity to the model, then the model too must be thought of as being endowed with change and body. For the animal is a substance endowed with soul, sensation and the power of move- 20 ment. It is hard to imagine, therefore, how that entity in the other-wordly realm could be an animal, or how he could be mortal, or how in general <such otherworldly entities> could be squared with naturally constituted ensouled things which are in motion, such as flesh, bone, man etc. Aristotle also shows that not all features can be separated from natural bodies by the fact that although the sciences do by nature separate, even so those more concerned with nature and motion (such as optics, harmonics and astronomy, which make use of 25 the senses and moving subjects in addition) cannot separate these aspects from matter and the natural world. For the science of optics deals with lines as does the geometer, and the science of harmonics deals with numerical ratios such as 9:8, 3:2, 4:3 and 2:1 as does arithmetic, and astronomy deals with spheres and circles as does the geometry of spheres. Even so, the former-mentioned of each pair of 30 sciences cannot separate these features from the moving and enmattered material in the way that the latter-mentioned ones do, defining them in their own terms and treating them as separate features. Yet the optical scientist gets his first principles from the geometer, the harmonic scientist from the arithmetician, and the astronomer from the geometrician of spheres. They are dependent on them in the way that the doctor is dependent on the natural scientist. It becomes 295,1 apparent that scientists engaged in optics, harmonics and astronomy cannot separate their subject matter from natural bodies when each of them is actively engaged in research and applies himself to his proper business, since each stands in an inverse relation to the more fundamental branch of science. 'For geometry deals with natural lines [as boundaries], but not *qua* natural, while optics [takes] mathemati- 5 cal lines (for they are merely length without breadth), but treats them *qua* natural, not *qua* mathematical.' The diagrams of optical science show that it deals with mathematical lines. If, then, it is not always possible to separate lines, numbers and circles although they are in definition separable, how much truer is this of natural entities and 10 compounds themselves, whose definitions are given as those of natural entities and compounds?

This was how Aristotle so convincingly countered the popular views about the forms, which are not conscious of the other world, but without even stripping enmattered forms completely of their matter[130] imagine that they pre-exist in the mind of the craftsman;[131] all entities in this world, both those with matter and those which share in the manner of the substantial being of things 'down here',[132] are considered simply, and they think that these are forms; yet they consider that the similarity of things 'down here' to things 'up there' not as that of image to model, but as absolute identity. For indeed if the enmattered man who enjoys perception and movement, is bodily and has a mortal existence is the man, then there would be no man 'up there'. But if things 'down here' are generated, and if whatever is generated must be so through a cause – and not a common cause but one which is distinct and productive of each individual item – then necessarily the causes of all things that are generated must pre-exist in the craftman's mind as distinguished according to their forms, just as in the mind of the builder there is one idea that will produce the wall and another that will produce the roof. When we apply the names of things 'down here' to things 'up there', I do not think it all absurd if we call them by the same name as being their distinct causes and having the same definition 'up there' as do the end products 'down here'.[133] For Aristotle himself said that there is a twofold order, the one in the world, the other in the cause of the world, just as there is one discipline in the army and another in the mind of the general.[134] Yet if anyone defined the order of things 'down here' just as they are 'down here', the definition would not exactly match the order in the mind, nor consequently would the name. For if there was some definition of the image Socrates as an image, it would not exactly match the model. Yet if one were to include in a definition the similarity of the image to the model, for example snubnosedness, popeyedness or baldpatedness, ignoring the material substrate, not even then would the definition exactly match Socrates, although it will be like his definition. For not even the snubnosedness in the image will be the same as that in Socrates, but will be like it and an image of it. But neither Aristotle nor his followers dispute that natural things have pre-existing causes and definitions from which they stem. But those who posit the Forms claim not only that they are causes of the things in this world, but also the models, since things 'down here' owe their existence to their similarity to them. Therefore they use the same names for both, which would be nonsense if the natural forms were not also mentally separated.

It is therefore worth remarking that Aristotle too knew the compound of matter and form as one thing, and the form itself as another. In point of fact he claims on each occasion to give one definition of the matter, another of the form and another of the compound. Moreover

in this passage he is envisaging the form *per se*, when he says that the nature is the form and the shape. For on this occasion he does not mean the compound when he says: 'What then is it that is growing?[135]
15 Not what it started growing from, but what it is growing into. The form therefore is its nature.' It is clear that the definition of the form in this guise is not given with respect to the compound, nor does it include the matter, but results from an abstraction from the compound. It is in no way absurd in such cases to analyse the definition of the compound into the definitions of its simple constituents, as he
20 himself says elsewhere.[136] So it is possible to separate mentally not only mathematical forms, but also natural ones, since it is possible to give the individual definitions of the forms without their matter. But this is not the mentally conceived model; rather the definition according to the form, when separated, can to a certain degree be made to match the model if one understands it is as being like, but not the same as, the model, and <as the definition> of an image likened to its model. Plato calls the model 'a living thing', saying in
25 the *Timaeus*:[137] 'In so far, then, as Reason sees the Forms present in what is the living thing ….' It is clear that this is no natural living thing, but the intelligible model of the natural living thing. Similarly Socrates is not the picture of Socrates, but he is the model of the picture; he is properly Socrates because the likeness belongs to the image not according to the matter but according to the form which is
30 mentally separated from the matter.

I think that it is possible to use Aristotle's hypotheses to show that the causes of forms in this world are distinct from them and are models for them. We say that natural entities exist as a result of the
297,1 participation of matter and form in that the matter participates in the form according to the participation within itself. All participation is intermediate between that which participates and that in which it participates, and is something conceded, by that which is participated in, to that which participates. The participation is the same in kind as that which is participated in, as the heat and burning in a body that is being heated by a fire is like in kind to the heating agent. Each
5 thing gives a share of what it has to whatever participates in it. Consequently, since the form in this world consists in participation in the separable form, it is like the separable form. What is like the separable form, coming-to-be in relation to it and being an image of it – that has the separable form as a model that belongs to it. And Aristotle himself calls the separable causes forms. Since things 'down here' are participations in things 'up there', they would be necessarily
10 like them, and it is perhaps hardly surprising that they share the same names.[138]

However it may perhaps not be conceded that the forms 'down here' are participations, but claimed that they are antecedent substances.[139] For 'up there' there is no man or horse, but causes of them subsisting in other forms on the one hand, and productive of man and horse on the other; for example, man comes from God, and the changeable comes from the unchanging – in neither case is there a similarity of form. People who make this claim should be asked 15 whether they find a place 'up there' for beauty, goodness, substance, life, knowledge, actuality, number – or anything of such sublimity. Aristotle makes it clear that he agrees that these do exist 'up there'; he says[140] that is a desirable state of affairs for everything and that mind is actuality in substance, bringing under one heading mind, substance and life, attributing to them beauty and goodness, and 20 saying that the immovable causes are the same in number as the heavenly bodies. Even so, Aristotle is clearly unhappy with those notions of the forms which, together with the names taken from 'down here', also drag along with them the definitions which include the physical and material elements of the things 'down here'. That is why he is unhappy with some of the names, although he does not think it wrong to apply other names – the more sublime such as beauty, 25 goodness, essence, life, mind and actuality – to the world 'up there'. If, then, someone were to say that those forms do exist 'up there' but are not of the same kind as those 'down here', we too would agree, but require him to say the same of man, horse and suchlike. For we too do not think that the man 'up there' is bodily since not even the man in his natural definition is bodily, but we do consider that there is 30 some likeness between the bodily and the unembodied man, just as there is some similarity between bodily and unembodied beauty. In general, if the forms 'down here' are generated, and if everything that is generated must of necessity be generated by some cause which has previously acquired the definition of what is generated (so that it should not be an irrational or indefinite generation), and if what is generated bears a resemblance to the definition, this cause would be 35 the model of whatever is generated.

Furthermore, if prior to all things that are changing eternally 298,1 there must be some unmoved cause of the uninterrupted change – not just a common cause, but a distinct one for each instance (as Aristotle himself demonstrates both in this treatise and in *On the Heavens*[141]) – and if each of the forms 'down here' admits change in each individual aspect, then it is clear that there would be a pre- 5 existing unchanging distinct cause of each form, for example a standing man, so that that which is changing too may have uninterrupted change. Moreover, Aristotle on several occasions likens nature to art, for example a little above this passage where he says:[142] 'For just as that which is according to art and artificial is said to be art, so that which is according to nature and natural is said to be nature.' If, then, art produces things according to descriptions, making the products 10

like the descriptions of the artefacts, it is clear that nature too produces things in this way, and that is why it produces nothing to no purpose. If, therefore, nature is not a first cause, and there is something else which is the cause of all things as nature is the cause of natural things, it is clear that this too starts with the descriptions and causes of everything that comes-to-be. It makes these products
15 like their descriptions. Thus it is possible to mention these things and many others, which, even on Aristotle's assumptions, show that the paradigmatic cause[143] pre-exists whatever comes-to-be.

There are many other elegant arguments to show that any form that has a distinct existence is an ungenerated and unchanging source. It would not be the form which is in matter in a primary way,
20 a participation and an affection of the matter, nor would it be the form that is in the nature as in a substrate, having substantial existence, nor would it be the form in the soul. For even if it is separable, even so, being there as something self-moved it has some duality,[144] and is no more mover than moved, and so it is not so much a principle as a derivative of a principle. The principle, therefore, must be unmoved, ungenerated and productive in only one way, and not something that
25 also comes-to-be. Such are the forms that are distinct in mind in a primary manner, unmoved, ungenerated and truly existent, which are said to be the causes of what comes-to-be, to which things that do come-to-be are likened.[145] If, however, there are names which are not of the forms 'down here' but of compounds,[146] we should not transfer the likenesses of them beyond this world, because although the particular characteristics which come from above right down to the
30 lowest level remain the same as they proceed downwards, nevertheless the modes of existence – for example 'the unmoved' or 'the self-moved' or 'the enmattered' – remain at their appropriate levels and are not transferred. If, then, there is some name which is of the form with matter, we should not seek a model for this. For models are of forms, not of matter. For snubnosedness carries along with it
299,1 the nose, and since nose too is a form, it would not be surprising if it were to have as a model one of the things that are by nature, but not a cause for what is snubnosed only *per accidens*. But if a form is taken together with the matter, and if the name and the concept relating to the name carry along with them the matter with the form, we should not seek a model for this – a position which I think that
5 Aristotle is very much at variance with. Consequently, if one is able to conceive of the form of the animal by itself, separating from the matter the particular characteristic which participates, I do not think it anything strange, even according to Aristotle, to assume as a preliminary a description which allows matter to take on its likeness. If in the case of a man there is some form 'man' in which the matter participates, there is nothing to prevent us from ascribing its likeness

to the world 'up there'. But if the name and the thing can only exist with the matter, as in the case of snubnosedness, then the person trying to ascribe that to the world 'up there' must undergo Aristotle's scrutiny. 10

> **194a12** Since nature can be spoken of in two ways – form and matter – we should make our inquiry as if we were asking what snubnosedness is. Such things cannot be thought of without their matter, nor in terms of their matter.

Having said that some of the mathematical sciences separate boundaries and properties from their matter and consider them in this light (for example, geometry deals with natural line but not *qua* natural), while others such as optics make use of what has been separated from its matter (for example mathematical line, but *qua* natural, not mathematical), he naturally proceeds to ask how natural science recognises what is natural. He says that it recognises it as what is a compound of matter and form, but not in terms of its matter but of its form; for example, the person recognising what is snub recognises it together with the nose, but in terms of concavity. Quite rightly so; for the natural scientist is a student of nature. But nature is spoken of in two ways – according to matter and according to form – and it is the latter which is the stricter. Therefore he quite reasonably recognises the compound in terms of its form. And since matter is knowable by analogy, while form is knowable by experience,[147] our knowledge of the compound is made complete according to the form, as you would expect. 15 20 25

> **194a15** In fact one might feel that there is an ambiguity here too. Since there are two sorts of nature, which one is the concern of the natural scientist? Or is he concerned with what is compounded of the two? If so, he must be concerned with the two components, and then will have to ask whether each falls under the same or a different study. To anyone paying attention to the ancients, it might seem that he is concerned with the matter, since Empedocles and Democritus barely touched on form and essence. But if art imitates nature, and it is the concern of the same science to know, up to a point, both the form and the matter (as it is the concern of the doctor to know about health and also about the bile and phlegm on which health depends; as too it is the concern of the builder to know about the form of the house and its material constituents, that they consist of bricks and timber; and so on), then it is the concern of the natural scientist to know both sorts of nature.

Furthermore, the same science must be concerned with the goal or end and the means to it. Nature is an end and goal. For where, in the case of continuous change, there is some end in sight for the change, that is the final point and the goal. That is why it was rather laughable when the poet was led to write: 'He has reached the end for which he was born.' For the final point should not be just anything, but only the best.

30 He has said that nature can be spoken of in two senses and that it is 300,1 the concern of the natural scientist to discuss both, but in terms of the more important of the two; and he has declared this without any formal proof as it might appear on a superficial reading – since he himself did indicate the proof (for if nature is spoken of in two ways and the one is more important and more knowable than the other, then it would be the concern of the natural scientist to discuss both ways in terms of the more important – except that this has been 5 stated without proof);[148] he now says that it is a problem, if the elemental natures are two in number, to decide which of the two the natural scientist should be concerned with. He pushes the enquiry forward by means of a sort of division: Is the natural scientist concerned with just one of the two, both of them or the composite of them? If this last, then it is clear that he must be concerned also with the two components, for it is impossible to know the compound 10 without knowing the elements. If he is concerned with each of the two components, matter and form, is it the concern of the same science to know each of the two, or is it the concern of two separate sciences? But in making such a division he passed over the section <of the division> 'the two components', intending to deal with it along with the compound. So in this way having presented the question as a division, he shows that the ancient natural scientists busied themselves with the enquiry about the one aspect, the material one; 15 Empedocles and Democritus[149] scarcely touched on nature according to the form. For Empedocles posited Strife and Love as causes producing the form among his first principles, while Democritus posited shape, position and arrangement, defining the form, I suppose, in terms of the proportion according to which they rendered each thing; for it is according to a certain proportion that Empedocles renders flesh, bone and everything else.[150] In the first book of his 20 *Physics* he says:

> The grateful Earth took into its well-made crucible two of the eight parts of bright Nestis, and four of Hephaestus. They became white bones, joined together by the divine bonding of Harmony.

By that he means 'by divine causes, especially Love or Harmony', for it is by her bonding that they are joined together. 25

Alexander writes: 'Aristotle failed to mention Anaxagoras,[151] although he placed mind among his first principles, perhaps because he makes no use of it in the process of generation.'[152] But it is clear that <Anaxagoras> does, since he says that generation is nothing other than a separation, and separation happens as a result of 30 movement, and mind is the cause of movement. Anaxagoras says:[153] 'When mind began to cause movement, it began to be separated from the moving all, and however much mind set in motion, that much was separated. As things were set in motion and separated their revolution caused them to be separated much more.' The reason why 301,1 <Aristotle> did not mention Anaxagoras is that Anaxagoras did not say that mind was an enmattered form – the object of <Aristotle's> current enquiry – but a separating and organising cause, distinct from what it organises and being of a different order of being from what it is organising.[154] Anaxagoras says:[155] 'Mind is indefinite, 5 self-ruled, not mixed with anything but subsisting alone, by itself', and he gives the reason for this. Perhaps another reason why <Aristotle> failed to mention Anaxagoras is that according to him mind seems not to create but to separate out the forms that already exist. But it is clear that separation out from unitary Being, whereby 'all things are together',[156] is intellective production.[157] 10

Having said this <Aristotle> goes on to add arguments showing that the natural scientist must know both sorts of nature, the material and the formal – not just the material, as many natural scientists thought. The first argument starts from induction from the field of art: if art imitates nature, and if it belongs to the same science[158] to 15 know both form and matter, then it would belong to natural science to know both natures. He takes it as evident that art does imitate nature, and he uses the examples of the doctor and the builder to show that it belongs to the same science to know both matter and form. The second argument he employs is something like this: if it 20 belongs to the same science to know the end or goal and the means to it, as for example it belongs to the science of building to know the roof and the walls,[159] and if the nature and the form are an end and the goal, it is clear that it would belong to natural science to know not only the matter (as the ancient natural scientists said) but also the form. Or, to put it another way, not only the form, which is nature in the more important sense, but also the matter. That is Alexander's 25 understanding. But it is quite possible in either sense. But the proof that natural science is concerned 'also with the form' is correct when it is spoken in relation to natural scientists, while the proof that it is concerned 'also with matter' is correct in relation to what is dealt with subsequently.

30 The entire passage seems to be aimed at showing that natural science is concerned with knowledge of both form and matter. That the form and the formal nature is an end for the matter he more or less shows in the following manner: the matter and the natural change in the matter are continuous;[160] the form is a sort of end for the continuous change, for the sake of which the change occurs, ceasing when it reaches that end. The matter and its natural change, 35 then, has the form as its end. In fact it is the nature of the matter to exist for the sake of the shape,[161] and not *vice versa*. The substrate is 302,1 perfected and exists properly only when it gains the form. That continuous change is directed towards some end is clear from the fact that aimless change not directed towards some definite end is not continuous, but interrupted and irregular. Since not every limit and final point is the goal of something, but only the best is (for in the 5 case of aimless change and things being transformed because of a weakness in their nature there is an end in the sense of a final point, but this is still not the goal) the definition given is a reasonable one. Aristotle says that it was laughable for the poet[162] to say 'He has reached the end for which he was born'. For the end of life is a final point and a limit, but not an end in the sense of a goal. Birth and 10 growth do not occur for the sake of death, but for the form and the perfection of what is growing, towards which the growing thing moves continuously; when it reaches it, the forward process stops. But if nature looked towards death as an end, the animal would perish immediately upon reaching maturity. As it is, after achieving its proper stature it is often preserved. One could say the obvious, that 15 death is the end not only of development but also of natural life. But death is an end and a final point not in the sense of an aim, but as an inevitable attribute of this kind of substance. Alexander says that it would have been better to have written 'where, in the case of continuous change, there is some final point, that is an end and goal', since 20 not every final point is an end. Yet Aristotle himself agrees that not every final point is an end. Perhaps then there is no need to change the text. For where, he says, in the case of continuous change (i.e. change that has an aim and is directed towards some end), there is some end as limit and final point, that final point is an end as a goal 25 and not merely a final point, as would be the case with undetermined change.

194a33 For art makes its matter – sometimes absolutely, sometimes rendering it better to work with.

This seems to support the suggestion that the earlier statements demonstrate that natural science ought to be concerned with know- 30 ledge of the matter. Perhaps all that has been said, as I pointed out above, leads to this conclusion, that natural science is an investigation to do with the matter, the form and the compound according to the form. He takes it as agreed that it is to do with the form; he accordingly now maintains that it is the concern of the natural scientist to talk about matter as well, arguing again from the similarity of art, arguing *a fortiori*. For if art, in copying nature, not only knows the matter but also provides it – sometimes creating it absolutely as, for example, building creates the bricks, medicine the drugs, ceramics the clay, painting the mixture of colours, while sometimes rendering it better to work with as, for example, modelling softens the wax, sculpting melts the bronze and building shapes the bricks – it is all the more necessary that the natural scientist should at least be concerned with the matter that underlies natural things; for art in copying nature knows the matter in such a way as to be able even to provide it. Now if in productive arts the art that produces the form also produces the matter, it is also the case that in contemplative arts the knowledge which contemplates the enmattered form also contemplates the matter which underlies it. Perhaps the argument would be stronger if we said that if it is the nature rather than the art that makes the matter, then the natural scientist rather than the artist ought to know the matter. So in proposing to prove, from the fact that the investigation of the ends is part and parcel of the investigation of the means, that natural science is concerned with both the form and the matter, he proved (a) that the form is an end in the sense of a goal (because continuous change in the nature arrives at the form), and a little later on he mentions (b) that the matter is in relation to the form as to an end. At this point he shows that natural science is *a fortiori* concerned with matter too, as I have said, and he is not satisfied with the testimony of the ancients. 303,1 5 10 15 20

> **194a34** And we make use of everything, since it is there for our sake. For we too are, in a way, an end. For the term 'a goal' is used in two senses, as we stated in *On Philosophy*.

That the end and the means to it are part and parcel of each other he demonstrates from the fact that we are ourselves an end and make use of what we get for our own sake. And it is clear that what he has said about art and what he has said about ends demonstrate by extension what he has proposed. For doing and using are more than mere knowing. But in what sense are we ends, and in what sense do we make use of everything? The answer is that we use the term 'the end' in two senses, first as that *of* which the intention is (what people have more recently been calling an aim,[163] like health, which the doctor aims at), and secondly as that in which and *for* which it is completed, for example the person restored to health. It is in this 25 30

latter sense that we are an end. For the man is the end for the doctor
304,1 not in the sense of an aim, as health is an aim, but in the sense of that for which he wishes to produce health. The distinction is made by him in the *Nicomachean Ethics*, which he calls *On Philosophy*,[164] saying that the whole study of ethics is more truly to be called philosophy. The term 'the end', then, can be used in two senses, both
5 as that of which and as that *for* which; Aristotle has shown that the knowledge of the matter is part and parcel of the knowledge of the end in the former sense, and he now demonstrates that this is the case with the end in the latter sense, 'for whom', too. For being ends in this sense we not only know what is there for our sake, but we also
10 make use of it. He says that everything is there for our sake – not everything that exists whatsoever, but everything that exists for our well-being, such as the finished products of art, since we do make use of these. We are the ends of these products of art since they are there for us and exist in reference to us; they are inextricably linked to their ends. Not only do we know them, we also make use of them, which is something more. In general the person whose concern it is to know
15 the end is also concerned to know the means to it. So if it is the concern of the natural scientist to know the form as an end, it is also his concern to know the means to it, i.e. the matter. For if 'the matter is in the category of the relative' – i.e. relative to the form, as he goes on to say, and if at the same time knowledge is in the category of the relative, it is clear that whoever knows the one will also know the other of these too.

20 **194a36** There are, then, two arts which control and know the matter. There is the art which makes use of it, and the art which directs the production. Thus the art which uses is in a way directive. But they differ in that the one knows the form, and the other knows the matter. The helmsman knows and prescribes just what form the rudder is to take, while the boatbuilder knows what wood it is to be made out of, and by what methods it is to be shaped. In the field of art, then, we produce the matter for the sake of the job it is to do, while in the realm of nature the matter is already there. Furthermore matter is in the category of the relative, since different forms require different matter.

Having said that some of the arts not only know the matter, but even produce it, he now shows that although the directive arts seem to look in particular towards the form, not even they are unaware of the matter. For he says that there are two sorts of directive art. One uses
25 the matter – for example that of the helmsman, which knows just what sort of form the rudder is to take, prescribes the shape and size

and specifies the matter, that it must be wood which is neither too hard nor entirely pliable. The other art directs the individual productional skills, specifying the forms and the materials, as, for example, boatbuilding when spoken of inclusively, which directs the many skills involved in building a boat; it relates everything to the requirements of the whole, and specifies the type of wood from which the 30 rudder is to be made – that it must be perhaps cypress wood if it is to withstand the various strains which it is subjected to, as deter- 305,1 mined by the helmsman. But if the helmsman 'knows and prescribes just what form the rudder is to take', how can he be said to control the matter? So Alexander says: 'He too must know the matter when he is sailing the boat, which is not pure form, but is form together with a certain sort of matter'. Perhaps both the form and the matter 5 of the rudder comprise the matter for the user. That is why he says that in knowing and prescribing the form that the rudder is to take he controls the matter. It is possible that even if he does not know the immediate detail, that the rudder must be made of cypress wood, he does know in general that the wood of the rudder must be flexible but not too hard or too pliable. So knowing and prescribing this much 10 even he could reasonably said to be in control of the matter. He says that 'the art which uses is in a way directive' in so far as it too gives directives as to what the rudder is to be like. For the art which prescribes directs the art which produces. So even the person using it knows not only the requirements which are an end, but also knows the means to meeting those requirements – for example the matter 15 – and the nature of the end – i.e. the form. He knows the means to the end in more immediate detail.[165]

Having said that some arts also produce the matter he firmly establishes that just as in the field of art the matter is subordinate to the art, so in the case of things that exist rather than are made, for example in the realm of nature, the matter exists for the sake of the form. He may perhaps not mean that, in the field of art, the art 20 produces the matter, while in the realm of nature it does not (for it is the case that nature rather than art produces the matter), but that the scientist does not produce the matter, but in contemplating what is already in existence he himself relates it to the form like those who do produce it.

194b8 Furthermore matter is in the category of the relative, since different forms require different matter.[166]

From the fact that 'the matter is in the category of the relative' – i.e. 25 relative to the form – he shows that it is the concern of the same person to know both the form and the matter. He proves that 'matter is in the category of the relative' from the fact that a different matter

belongs to each form according to the appropriate relationship between the form and the matter which receives it; consequently, this particular matter would be the matter of this particular form, and this particular form would be the form of this particular matter. If this is the case, then the person who knows the one must necessarily
30 know the other. But how could 'different forms <require> different matter' if there is a common matter for all things? The answer is that it is clear that 'different' <refers to> 'proximate'. Just as each different instance of proximate matter is relative to its own form, so common matter relates to form in general. So the person who knows and understands the natural form *qua* natural will also know the matter which acts as substrate to such an enmattered form. Therefore
35 natural science is to do with both the form and the matter, and also, of course, with privation, since things that are in opposition to each other in this way belong to the same science.

306,1 **194b9** To what extent, then, must the natural scientist know the form and the essence? Just as the doctor must know the sinew, or the bronze-smith the bronze, up to what point[167] <will the natural scientist> take his knowledge? For each of these things exists for the sake of something. He should know about them in so far as they are separable in form but exist in matter. For both man and the sun generate man. It is, however, the task of first philosophy to define the character of the separable and the essence.

Having shown that it is the task of the natural scientist to know both the form and the matter, he quite reasonably chooses to ask next up to what point the natural scientist should enquire into each of these.
5 But he seems to have already decided about matter, that the natural scientist will enquire into it just up to the point where he refers the discussion of matter to the form.[168] So it remains for him to ask up to what point he should know the form and the essence. But having stated that the discussion about the form – but not in isolation from the matter – will be primary for the natural scientist, he now quite
10 reasonably raises the question of how far the knowledge of the form (clearly the natural form) is appropriate for the natural scientist *qua* natural scientist. For there are forms in a higher realm than the natural,[169] and it is not the concern of the natural scientist to talk about these. He says that the natural scientist *qua* natural scientist will talk about the natural form up to the same point that the bronze-smith and the doctor will talk about bronze and sinews respectively. Each of these two will talk up to a certain point – just
15 as far as each is useful for the appropriate art; neither of them will be worried about the first principles of the sinew or the bronze. The

natural scientist too, then, will talk about the natural form, up to the point where this is a substance. For each natural entity comes into being and exists for the sake of something, i.e. the end of the entity. By discovering this, he has come to the end[170] of his enquiry about natural forms, focusing his enquiry on these forms that are separable 20 in account but are enmattered. So the natural scientist will enquire not about any form, but about the enmattered form, and his enquiry will also proceed up to the point of asking for the sake of what it exists and comes-to-be.[171]

I think that the text reads better if we use an accent on the first syllable of the word *tinos* in the phrase 'up to what point?'[172] (for it means the same as 'to what extent?'), and also if we put an accent on 25 the first syllable of the word *tinos* in the phrase 'for the sake of what does each of these things exist?' (for the words up to the end of the phrase, i.e. up to 'for the sake of what?' show that this is what he means). In fact, if he were to have written '<the natural scientist will enquire only> up to the point of asking "for the sake of what?", for each of these things exists for the sake of something' (as Alexander says he did) we shall have to use an accent on the first syllable in the first case and on the second syllable in the second.

Having said that the natural scientist will be concerned with the 30 natural forms and will continue his discussion up to the end,[173] he shows that the natural forms are enmattered from the fact that their causes (both the immediate and the higher) are themselves enmattered. For the immediate generator of a man is a man, who is himself 307,1 enmattered; and the higher cause of both of them, which is productive of everything that comes-to-be according to nature, the visible sun, is itself an enmattered form.[174] Alexander says: 'In general that which comes-to-be by nature itself seems to be produced by what is the same as it in either species or genus, or, generally speaking, by activity. At any rate those things that are produced in this way by the sun are 5 produced by its activity; they are heated or cooled by its activity. It is clear that the things that are produced by art are also produced by an activity, but not a natural one.' Perhaps a man, fathered by a man, is produced by what is the same in species, and in so far as he is produced by the sun is produced by what is the same in genus – enmattered entity from enmattered entity. Alexander further infers 10 from this passage that according to Aristotle the coming-to-be of things in this world is dependent on divine body, and not unconnected with it.[175]

But in what sense, in reply to his question 'To what extent, then, must the natural scientist know the form and the essence?' does he say 'Just as the doctor must know the sinew, or the bronze-smith the bronze, only up to a certain point'?[176] For the bronze is the matter for 15 the bronze-smith and the sinew for the doctor. For it is with these

materials that they produce the forms specific to their art – the bronze-smith producing, for example, that of the vessel, perhaps, by means of the bronze, and the doctor producing that of health by dealing with the sinews. In what sense, then, does he say that the natural scientist enquires about the form up to the same point that
20 these practitioners enquire about the matter? Alexander takes note of this difficulty and adds another explanation, writing: 'Or else having first said that the natural scientist will talk about matter up to a certain point as do these practitioners (for in fact they do enquire up to a certain point about the matter that is available to them) he then immediately goes back to talking about the form, asking to what extent and about what kind of a form the natural scientist enquires. He enquires what goal each of the things that come-to-be comes-to-be
25 for, namely the natural form and the perfection of whatever comes-to-be by nature. He enquires therefore about the kind of form for the sake of which each of the things produced by nature comes-to-be, and about those forms which are separable from matter in account but have their existence in matter; he no longer directs his attention towards the bodiless and separable forms.'[177] The explanation would have been satisfactory if <Aristotle> had proposed that 'the natural
30 scientist enquires *about matter* up to certain point' and had gone on to say 'up to the same point that the bronze-smith enquires about bronzes, or the doctor about sinews'. As it is, having said 'To what extent, then, must the natural scientist know the form and the essence?' he adds 'Just as the doctor knows the sinew, or the bronze-smith knows the bronze'. And Alexander adds another explanation to meet such a difficulty to suit the reading '... the doctor knows the sinew, or the bronze-smith knows the bronze up to <knowing> what
35 they are for'. He says: 'The meaning would be that just as the doctor and the bronze-smith talk up to a certain point about the form and about that for the sake of which the matter, about which they are
308,1 concerned, exists (for they talk about the form for the sake of which the one concerns himself with sinews and flesh, and the other with bronze – about the form that is instantiated in these, and not about any other), just so the natural scientist too must talk about the form of which the matter is the matter and for the sake of which the matter
5 exists, i.e. the enmattered form. He will concern himself "about them in so far as they are separable in form but exist in matter", i.e. about such forms, and he will enquire about such forms.'[178] But I do not think that in saying this he explains to what extent the practitioners talk about the form for the sake of which the matter, which is their concern, exists – unless he is prepared to say to what extent they are
10 enmattered. For he said this more clearly in the case of the natural scientist.

Themistius[179] too seems to be of this opinion when he paraphrases this passage: 'About the form itself he asks "up to what point?", and the answer is "short of separating it from its matter; just as the doctor enquires about the sinew, so the natural scientist should enquire about the man".' Perhaps, then, the next statement 'each of these things exists for the sake of something' is not supposed to fit in with this.[180] For this does not refer to the knowledge of the form as something enmattered; rather the enquiry is about the form, to what extent the natural scientist must know the natural form, which is his concern. But he says that he should know about it to the same extent that the bronze-smith knows the bronze or the doctor the sinew, not presenting these as the matter of these arts, but as the forms with which the arts are concerned, just as natural science is concerned with the natural form. These arts have a practical concern, but the concern of natural science about its particular subject-matter is theoretical. So just as these arts know the subject-matter to do with their particular activity up to a certain point – the need which they present for reaching perfection, the very reason for their employment – just so natural science will know the natural form as far as knowing the need for the sake of which it comes-to-be and exists. For all knowledge which stems from the causes is most appropriate for a proper understanding, but that which stems from the final cause is the most important of all. Aristotle seems to be agreeing with Socrates in the *Phaedo* when he tells the natural scientist to press his enquiry into cause as far as the final cause. For in the *Phaedo*[181] Socrates says 'If anyone were to want to find the cause of each thing, how it comes-to-be, is destroyed and exists, he should discover this about it – how it is best for it to exist or to affect something else or to be affected itself'[182] – in other words he should discover for the sake of what each thing comes-to-be or exists. Having said that the natural scientist is concerned with what is separable in account but enmattered in reality, Aristotle rightly adds: 'It is, however, the task of first philosophy to define how the form which is separable in reality relates to the enmattered form, and what it is *per se*, and just what its essence is.' This he does in the *Metaphysics*.[183]

<CHAPTER 3>

194b16 Having made these distinctions we must now consider the causes. 309,1

The treatise is about natural principles and causes, which is why at the outset Aristotle himself said[184] 'We think that we know each thing when we discover its first causes and its first principles, and get as far as its elements.' He proposed to enquire first about the principles as elements, presented the opinions of the ancients about them, and

gave his own thoughts on the question, showing that even the ancient natural scientists agreed with him to a certain degree in that they too posited the opposites as principles. So he suggested matter, form and privation as elemental principles of natural entities; he demon-
10 strated what nature is (that it is a principle from which change <originates>), and what the terms 'by nature' and 'according to nature' mean; he distinguished the different senses of the word 'nature' – showing that form and matter are nature, that the natural scientist must concern himself with both, that the answer to the question 'to what extent <must he know them>?' is 'as far as the final cause'; he then filled out the discussion about causes, since he spoke
15 both about matter and form as element and about the efficient <cause> – namely nature – and about the final, since much was said in the course of arranging the causes in relation to each other; taking up the argument he will proceed to enumerate them, adding the appropriate definition for each of them, bringing under scrutiny first the causes *per se*, and then *per accidens*.

20 Here he again reminds us of the need of a treatment of principles, effectively arguing as follows: The study of nature is theoretical, not practical in the way that the study of ethics is; the goal of theory is knowledge; that which has knowledge as its goal will start with an understanding of causes, to which it will refer each thing and so gain an understanding of it; thus discussion of causes is proper to the study
25 of nature, in order that we should be able to give the reasons for coming-to-be, passing away, and all other natural change – local change, qualitative alteration, perhaps even growth – the proper concerns of the natural scientist. Why each of the subjects of enquiry *vis-à-vis* these modifications is of such a character we discover by relating each to its causes and principles. For we shall think that we are giving the causes by showing that it exists because it is of
30 such-and-such <a material>, or because it is of such-and-such a kind, or because it is affected by such-and-such an agent or because it is better so.

> **194b23** In one way therefore 'cause' means that from which anything comes-to-be as from something inherent; for example, the bronze is inherent in the statue, the silver in the cup; this is true on a less specific level too.[185]

35 He says that there are four ways in all of defining the term 'cause',
310,1 and he starts with 'that from which anything comes-to-be as from something inherent', i.e. the material and the substrate, which seems to share with privation the fact that something comes-to-be from it, but differs from it in that it is something inherent. For whatever comes-to-be does so *from* privation in the sense of *following* the

privation, where the privation is lost,[186] but from matter as something inherent which changes from one disposition to another. Therefore the term 'that from which' is ambiguous. Matter differs from form, which is also inherent, because that which comes-to-be does not do so from the form (for the form is neither lost nor altered) but according to the form. He listed matter first among the causes because the earlier thinkers defined the causes of almost everything by relating them to matter, for they said that dice fall downwards because they are made of bronze, and that a cauldron lasts longer than a pot because the one is made of bronze and the other of earthenware. The reason why he listed matter first is that he was proceeding 'from the bottom upwards'. And not only is the proximate matter a cause of what comes-to-be, but 'this is true on a less specific level too'. For not only is this particular lump of bronze a cause of the statue, and this particular lump of silver of the cup, but bronze and silver are causes in a general way. And if these things are <reducible to> water,[187] then water, and at a higher level, body are causes in a general way.

> **194b26** In another way it is the form or the model; this is the definition of a thing's essence – also true on a less specific level. For example, the ratio of 2:1 is the cause of the octave, and in general number is a cause, as too are the parts of the definition.

The second cause he defines as 'formal'. For inherent health is a cause of being healthy, and <inherent> commensurability is a cause of being commensurable. But form is not only the immediately apparent shape, but is also to do with the account. But when he calls the form a model, he is not suggesting that it is some ideal self-subsisting substance to which the things in this world bear a likeness, as do those who posit 'the Forms'.

Alexander says: 'Things that are productive in nature do not first of all have a conception of what they are producing, and then produce it in such a way that one could say that according to Aristotle the conception is a model of what is produced, as is the case with the arts; rather it is the form which is instantiated in matter which he calls a model because nature produces whatever it produces by aiming at this. This is clear from the fact that when it has been produced, nature ceases producing it, since the form is something defined and, as it were, a target set up at which nature aims, which is the reason for its being called a model.[188] But the end and the model', says Alexander, 'do not have the same significance in the case of everything that produces for the sake of something. On the one hand, in the case of things that produce according to choice, art and reason, the end must be that for the sake of which everything else comes-to-be; it must first be conceived in the mind of the producer and be set up as a target and

35 model for what is to be; on the other hand, in the case of things that come-to-be by nature, this is not so. For nature does not work by 311,1 choice or by any reason within it, for nature', he says, 'is an irrational power. Rather the first principle is lodged in the matter which receives both it and what is to come into being because of and from it; that first principle, itself lodged, itself produced that defined entity of which it is itself productive, while what results from this is 5 something different. For each of them[189] is productive of, and causes change in, what comes after it – if nothing stops it. This process continues up to an end, the attainment of the natural form, of which the beginning was the principle lodged in the matter – just as in the case of marionettes; when the operator starts the movement in the first doll this becomes the cause of movement to the next, and this to 10 the one after it, until the movement passes through them all – unless something prevents it – not according to any reason or choice in themselves when the one before moves the one after.[190] In this way too the nature-cum-power is lodged together in the seed, and becoming present in the appropriate matter and having the ability to change it, does change it in so far as it is naturally constituted to make changes, and <matter> to be changed. The power which is engendered 15 from the first change produces in its turn a second change, and keeps its force until it has produced something like that from which it started when it was lodged in matter,[191] and something the same either in species or in genus, as is the case with animals that are produced by dissimilar begetters, such as mules, which are the same in genus as their begetters. This progression proceeds according to stages and regularity until what is coming-to-be is perfected according to the form, if nothing prevents it.[192] This does not happen 20 according to any reason or choice in the agents of change and production, as has been said. But nature does not, on account of this not being the case, produce at random and not for the sake of something. "For the sake of something" is not the term applied <only> to coming-to-be based on reason and choice, but everything that comes-to-be according to some regularity and because of something else does so for the sake of something. This is the case whether there is choice and 25 reason or whether there is no reason, as we say is the case with nature. For even if natural things are, through material necessities,[193] accompanied by what does not have reference to anything (as is said to be the case with hair in the arm-pits, unless even this does have some use), one should not represent nature for this reason as not acting for the sake of something. The form, then, is a model in that nature has nodded in its direction not through choice, but more 30 like a marionette. In the case of natural things,' he says, 'the form of the producer is the same as the form or the genus of the thing produced and it too would be a model. In general since those who

produce something according to a model produce it according to something determined, and since it is special to that which is produced according to a model to be produced according to something that is both determined and like it, and if something is produced 35 according to something determined and like it, then it would be produced according to a model. This is how the products of nature come-to-be. Therefore they are produced according to a model.'

Since this is what Alexander says, it is clear that Aristotle calls 312,1 the form a model not in the sense of a 'Form'.[194] He takes the enmattered form as a cause in so far as the essence[195] embraces the compound object. He shows this in what follows by counting the matter and the form (in this sense) among the elements. But if he talks about the elemental form and calls this a model, how is it 5 possible now to understand the form of the producer as a model? For even if it is in some sense the same as what is produced, in that a man begets a man, even so he now calls the product and not the producer a model. For the product is the elemental form in that the essence embraces the compound. But the natural object is produced according to something determined and like it. For if it were produced 10 according to the producer, the producer would be the model. First, as I said, Aristotle did not name that, but the product, as the model. Secondly, how could it still be possible to claim that natural things are not produced according to some model, since we agree that the producer is the model for the product in the sense that it is produced according to it as something determined and already there similar to the product? But if the form is said to be a model because nature 15 makes everything by aiming at it, then for this reason it would be an end and not a model. For the object aimed at and that for the sake of which the product is produced would be an end and not a model.[196]

Furthermore, what was said about the coming-to-be of natural things seems to me to deserve close examination. For if what was initially sown itself produced that of which it was itself productive,[197] 20 and if what was produced from it was something different, and if each of them was productive of what followed after and able to cause change until the attainment of the end and the natural form of which that which was initially sown in the matter was the starting point, then – if the process of coming-to-be reached its conclusion in this way – first, if the grain produced the shoot, and the shoot produced the stalk, and the stalk produced the ear, how could it be the case 25 that the less complete produced the more complete? For the shoot is less complete than the stalk, and the stalk less complete than the ear. Secondly, we cannot say what the cause of the whole is. If the form is single,[198] then there needs to be a single cause prior to those of the parts. In general if the subsequent products come-to-be when there is an alteration in the earlier ones, the former products would be

30 material rather than productive. For the productive cause never becomes the completed product by being altered. Furthermore, if things are not randomly produced, but only from what has the <appropriate> potential, it is clear that the shoot is the stalk in potential. Everything that exists in potential is material, not productive, so that the shoot could reasonably be called the matter of the stalk, but in no way its productive <or efficient> cause. Secondly, 35 everything that comes-to-be in actuality from what exists in potentiality is brought to actuality by an actuality. If, then, the shoot is not the stalk in actuality, it could not produce the stalk in actuality. In 313,1 general if the nature of each thing, being a source and cause of change, is productive of its own substrate and no other, then it is clear that the nature of the seed will produce the seed, and the nature of the man will produce the man. How, then, could the nature which is in the seed but which is not yet the nature of the man properly produce 5 a man before the man comes-to-be? The answer,[199] as I said before, is that as soon as it itself comes-to-be it begins to produce, because it is naturally better so and because it is a sort of life awakening and rising up to the form, since the seed of the male and <the input> of the female[200] have as their nature the alteration of the seed which finds its natural completion in the animal. The cause which is properly and 10 immediately productive in the case of animals is the maternal and the paternal nature, and in the case of plants it is the nature of the grain and of the earth; the form pre-exists in actuality in the father, the mother and in the rational principles[201] subsisting in actuality in the earth, according to which the potential is brought to actuality. If the nature of what comes-to-be is said to be productive in this way, it would be in this way that it is productive – in the sense that it itself 15 comes-to-be. But what is productive in the proper sense is the nature of anything that exists in actuality. For nature is the begetter of its like, for the sake of which it anticipates all the stages in between, even if the nature which is coming to completion and being is altered in the process of producing, although it maintains a thread of continuity until the end, until it attains that end and ceases its production. 20 Even Aristotle says[202] that what exists by nature is whatever moves continuously from some source within itself and arrives at some end in which the movement ceases. So that in man the rational principle of the parent has been taken forward, and it is according to that that what comes-to-be is produced; the father provides the beginning and the movement right up to the completion by means of the seed (just as in the case of the marionettes the operator provides the beginning 25 of the movement and the impulse towards its completion), according to the rational principle, pre-existing within him, of the whole ordered process of change, while the maternal nature is more immediately productive of the form. Why, then, does he say that nature is an

irrational power although it acts for the sake of some end, proceeds in an ordered way according to stages and determined measures? The answer is that the productive rational principle is twofold, one producing in a cognisant manner, which the interpreter[203] sees as reason alone, the other without cognition and self-contemplation,[204] but still producing in an ordered and determined manner for the sake of some prior end. Just as the noncognisant is irrational as compared to cognisant reason, so that which produces in a random and disorderly manner is irrational as compared to that which produces in an ordered and determined manner for the sake of something. Therefore just as what comes-to-be by nature does so according to a rational principle of this nature, so it does so according to a model which is not established as something known by the producer, but because the producer makes the product like itself by being, not by choosing, just as the signet-ring makes the impression. Aristotle himself in the second book of *On Coming-to-be and Passing Away*,[205] alluding to Empedocles, agrees that nature is a rational principle. But in what way is there an order and a determined end in the production without knowledge in the producer? For in fact natural things have the sort of existence that allows them, without any knowledge, to preserve, merely by being, the order and the thread of consistency in the process of their production and to reach a determined end, just as is the case with the movement of marionettes. But unless such natural things were the product of chance – which is absurd – either such things gave themselves existence, or they came into being from something different. Whatever the case, such existence occurred through being entirely known; for that which calculated its power against the product produced it, by knowing both.[206] Therefore it is sensible to say rather that nature is co-responsible, and that the proximate causes of things that come-to-be and pass away are the movements of the heavens, according to which things on this earth are modified; and that higher up are the psychic principles of the movements, and even higher than these are the intelligible forms, from which in the first instance the shining forth of the forms is produced in all things according to the suitability of the recipients.

30

35

314,1

5

10

Aristotle calls the enmattered form a model equally as a target for nature, at which it aims not by knowing but by being, so producing everything, and equally as a model that is produced for art, since he does not want natural things to be produced according to some model, while he says that artefacts do need some model. Having said a lot by way of introduction about the fact that art imitates nature, he now quite reasonably reminds us that the natural form is the model for art. It is clear that since intellect is in the proper sense the cause which produces in accordance with the forms within itself, making its products like them, the forms in mind would properly be called

15

20

models. But enough of that. He said that the definition of the essence is the account. The definition is indicative of the form.[207]

25 **194b27** ... also true on a less specific level. For example, the ratio of 2:1 is the cause of the octave, and in general number is a cause, as too are the parts of the definition.[208]

Just as he said that in the case of the material cause the cause was not only the immediate matter, but also that 'this is true on a less specific level too', so in the case of the formal cause. Since the concordance of the octave lies in the double ratio, the octave is one species of the double, of which the double is the genus. And number 30 in turn is the genus of this. So these, he says, would be the formal causes as being involved in the specific form. Similarly the parts which are implied in the definition of the specific form – these too would be formal causes since they fill out the specific form. If the <definition> 'rational mortal animal' is the formal cause of man, then each of the parts of the definition would be a contributory cause 35 according to the same mode of causality as the whole, namely the 315,1 formal mode. So he quite reasonably specified the parts of the definition, not those of the specific form, since each of the parts of the definition pertains to the whole form.

194b29 Then there is that from which comes the primary principle of change and its cessation; for example, the person who has deliberated is a cause; the father is the cause of the child; and in general the producer is the cause of the product 5 and the changer is the cause of what is changed.

He presents as the third of the causes the productive <or efficient>[209] cause, because what first meets the eye is the product and its elements, and then, having discovered that it is a product we go on to seek the productive cause, then the reason why the producer produces and the product is produced. He calls the producer 'the 10 primary principle of change and its cessation' because he wants the productive cause (in the strict sense of the term) to be separate and distinct from its product. For the inherent cause, such as the form and the nature consisting in the formal principle, is contained <in the product>. We should remember that Alexander, commenting on this passage, agrees that nature is not a productive cause in the strict sense, but is rather a formal cause since it is not foremost among the 15 producers. Instruments appear to be responsible for movement, but not even these are efficient in the strict sense, since they do not move in a primary sense but because they themselves are moved. We must also remember that Alexander agrees that even the instrument is an agent in a certain sense; even if it is not efficient in the strict sense, it is at least in itself instrumental. Since of the things that come-to-be some are moved while others are brought to a stop, that which moves the former and that which stops the latter would be efficient causes, the one of movement and the other of its cessation.[210] That is why he said 'the principle of change,' adding 'primary' and 'and of its cessation'. Thus the person who has planned something good or bad would be an efficient cause in a stricter sense than the doer, since the primary principle of change lies in the person deliberating. He then generalises as he did in the case of the material and the formal causes. For the father and the child <are subsumed> under the producer and the product, and these in turn under the changer and what is changed. For production seems to be to do with substances, while change is said to occur in the case of the movement of other things also.

20

25

> **194b32** Then there is the end in view; this is the goal, as, for example, health is the goal of walking. 'Why does anyone walk?' we ask; 'In order to be healthy' we say, and think that in doing so we have given the cause. This is true of whatever happens in the attainment of some end when something else initiates the process; for example, slimming, purging, medicines and surgical instruments are all for the sake of health – all these are for an end, and only differ from each other in that the two former are activities and the two latter are tools.

The fourth cause he presents is the final cause. That is the one that supplies the answer to the question 'For what purpose does any particular thing come-to-be?' For when we are asked 'Why does anyone walk?' we reply 'In order to be healthy'. And in making this reply we think we have given a sufficient cause (for health is the cause of walking, or of taking a medicine) not in the same sense as any of the other causes given above, but because one walks or takes medicine for the sake of health. When he mentioned one item, walking – the cause of which, in the sense of an end, is health – he indicated that he included also all the stages between the action of the efficient cause and the attainment of the end, for example, the action of the doctor and the attainment of health. That is to say the medicine, the taking of the medicine, the purging and the slimming are done with the end in view and for the sake of that end. And the end in view is the cause of their happening. These things differ from each other in that some of them are activities (the walking and the purging) while others are tools (the medicine and the scalpel); but health would be the final cause of them all. They themselves, as Alexander says, are the efficient causes of health; but perhaps it would be better to call

30

35

316,1

5

10 the latter 'instrumental',[211] since he himself agrees that what are not the primary causes of change are instrumental and not efficient in the strict sense. In this case the doctor is the primary cause of change in respect of health, and the intermediary causes are not primary. The treatment and the intermediary stages are causes of health, the former as efficient, the latter as instrumental. Health is the final cause of the treatment and the intermediary stages. They could also, 15 I imagine, be called the material causes of health.[212] Alexander, I think, called them efficient with an eye to what Aristotle says a little later on, where he claims that working is the efficient cause of well-being. They are said to be healthy because they are in an ordered relation to each other. Alexander says: 'Aristotle uses them to show the ambiguity of the term "healthy"; for all of them health is the cause, 20 in the sense of an end, of their being just what they are.'

195a3 Causes can be designated in more or less this number of ways.

The phrase 'more or less' is added because although there are just this number of causes in the strict sense, there are many incidental types of cause, as he will say, or else because he has spoken cau-25 tiously, since Plato numbered[213] the paradigmatic cause along with the causes in the strict sense, the efficient and the final, and the instrumental along with joint causes,[214] the material and the formal. If there is such an extensive hierarchy among the causes, with the ones that exist by nature – the efficient and the final – being the primary causes, those in the strict sense and matter and form being rather contributory causes, then it is quite reasonable to place causes among the things that can be referred to in many ways and not among 30 the things that are divisions of a single genus. But that the modes of causation are just so many – no more, no less – one could perhaps work out from a division, declaring no more than that causes are the reasons why an entity is what it is or becomes what it becomes, and with which we answer the question 'Why?'.[215] Natural entities, which are compounds of matter and form, are what they are and become 35 what they become either because of what brings them to completion or because of what bestows something on them in some way from 317,1 outside. Matter and form bring them to completion, and because of them they are enmattered and natural. When we are asked why it is that things in this world have extension while intelligible beings are not extended, we say it is because they are enmattered. When we are asked why the heavens are moved so easily, we say that it is because they are spherical and 'turn on the smallest pivot', as Plato puts it.[216]

5 These are the answers to questions about causes from the point of view of matter and form, to which all those things that bring the

compound to completion in terms of elements are related. But since all natural things come-to-be, even according to Aristotle, and since everything that comes-to-be has a cause for its coming-to-be, then in the case of things that come-to-be the producer must be the efficient cause, the origin of the primary principle of change. But the principle of change is twofold, the self-changing and the unchanged. That 10 which is changed by something else could clearly not be a principle of change if we take 'changed by something else' as one event preceding another. But the self-changed could be a principle of change in that it has the agent of change within itself, although it is not a principle in the strict sense, because it remains the same and yet is changed. Being the principle of change in the strict sense means being 15 an unchanged changer, so that the efficient cause in the strictest sense of things that come-to-be would be that which is unchanged, eternal and always remaining enduringly the same.[217] Such is august intellect. Below intellect comes soul. For even if soul is changed, it has the agent of change within itself. That is why Aristotle prefers to call it unchanged,[218] since he thinks that only those things that are altered in body are changed. Since of natural things those that 20 come-to-be and pass away come-to-be because of the proximate agency of what is eternal and in orbit ('for both man and the sun generate man'), it is clear that the producer in the strict sense causes change not by approaching what comes-to-be and passes away directly, but through everlasting intermediaries.[219] In this way it becomes clear to us what the instrumental cause is, that which is both changed by something else and itself changes another thing. This is evident in the production of artefacts; the adze is a contributory cause 25 in that it both undergoes and produces change. This is its whole and partial nature, as Alexander admits when he cleverly remarks that the efficient cause in the strict sense must be separate and distinct.

Again, since the natural generated form consists in participation of form residing in matter, and since all participation consists in an imaging of what is participated in, and since this imaging is the 30 means by which that which is participated in is imaged, and is the image to which that which participates is likened,[220] there must also absolutely be a paradigmatic cause of enmattered things. That which produces does so either at random and by chance, or else it looks to some prior target. An end for the production is established, for the sake of which the producer produces and the product is produced. If the producer in the primary and strict sense produces at random and 35 by chance, what would be the producer of those things that produce for the best?[221] The primary agent, then, must act for the sake of 318,1 something, and his end must be the goal. In this way the final cause is revealed to us from the primary producer in that it is established as an objective for the other efficient causes. It is clear that these are

causes of what comes-to-be. That they are the only causes must be
5 carefully considered from the division. For that which comes-to-be, which is, as we claim, natural and compound, is a substrate and is in a substrate, and is nothing else besides these, and it has its being either from itself or from somewhere else. If, therefore, it is impossible for what comes-to-be in the divided nature of time and is corporeal and in extension to have its being from itself, then it must have some other efficient cause beyond itself. And this is either changed or
10 unchanged. If the former, it is changed either by itself or by something else. But that which is changed by something else is not a primary agent of change, and that which is changed by itself either changes in part of itself and is changed in another, or it changes and is changed in its entire self, like the soul according to Plato. This is a principle of change and coming-to-be on the one hand, but on the other it does not provide uninterrupted change in that it is itself changed, nor is
15 it fully primary in that it has a sort of duality between changing and being changed. Therefore that which is unchanged, yet is productive and a cause of change in the strict sense, is primary, subsisting eternally.[222] That which comes-to-be and is changed in time by what is eternal and unchanged cannot come-to-be and cannot be changed by it without an intermediary (for that <primary cause> is productive, and causes change in things that are everlasting), but what is
20 changed everlastingly by it according to its own different dispositions is the instrumental cause of change in what comes-to-be and is changed in time, because it changes and is itself changed. If the enmattered form is either primary or derives from the primary and is related to it, but if none of the enmattered forms are primary in that they are participations, then there is something primary after which they are imaged. And if what produces in the strict sense produces either at random or by chance, or else by looking to some determined target, and the former is an impossibility, then there
25 must eternally be some end and some goal.

195a4 Since causes can be spoken of in many ways, it happens that there can be many causes of the same thing, and not *per accidens*. For example, both the sculpting and the bronze are causes of the statue not in any other respect, but *qua* statue – but not in the same way; the one is the material, the other is the source of change. Some things are the causes of each other; for example, work is the cause of well-being, and *vice versa*, but not in the same way: one is a cause as an end, the other as a source of change.

Further, the same thing can be the cause of opposites. That which, by its presence, is the cause of something else, is sometimes blamed for being the cause of the opposite by its absence; for example, the absence of the helmsman is blamed for the capsizing of the vessel; his presence would have been the cause of a safe voyage.

He has presented many distinctions among causes; but so that no one should think that there are different causes of different things (for example a material cause for one thing, and a formal or efficient for 30 another), he rightly indicated that as a result of what has been said 'it happens that there can be many causes of the same thing'. Having said that this 'happens', so that we should not imagine that this is sometimes the case, sometimes not, he adds 'and not *per accidens*', but in so far as the thing that comes-to-be is just what it is. For in so far as it is natural or artificial, it has matter, form, an efficient and a final cause. The words 'not in any other respect, but *qua* statue' 35 were said instead of 'not *per accidens* but *per se*'. For if it is *per se*, it 319,1 is obviously not *per accidens*, since it is bound to be the one or the other. He further says that not only are these the causes of what comes-to-be, but also that the efficient and final can be causes of each other. For work is the cause of well-being as an efficient cause, while well-being is the cause of work as a final cause, since work is done for 5 its sake. And perhaps well-being is also the efficient cause of work. For the fit man can work and be active. Having said that all things[223] can be the causes of the same thing and of each other, he adds 'the same thing can be the cause of opposites', but not in the same way, but in opposite ways. That which is the cause, by its presence, of a safe arrival for something – as the helmsman is for the ship – by its absence can be the cause of destruction.[224] It is clear that being 10 present he is a cause *per se*, and that being absent he is a cause *per accidens*. This is how he spoke in the first book,[225] saying that the form is the cause of coming-to-be by its presence, and of passing away by its absence.

195a15 All the causes we have mentioned fall into four very obvious classes. Letters are the causes of syllables, materials are the cause of artefacts, fire and the like of bodies, parts of the whole, premisses of conclusions as 'that out of which'. Some of these are the substrate (as are parts of the whole), while others – the whole, the composition and the form – are the essence. 15

In his account of causes he presented not only the genera of causes as causes, but also what lies between the agent and the product, i.e. the end; but he quite reasonably says by way of summary that the types of cause are four, to which all the ones that have been discussed can be reduced. Having proposed to enumerate them concisely, and

20 having first presented the material cause he wants to elucidate it and explain the distinctions by exemplification. For the sake of greater clarity he mentions also those things of which it is a substrate, naming 'the letters of syllables, the material of artefacts' and so on. Among these the letters act as substrate, as does the material of 25 artefacts, such as wood, wax, olive oil for medicines, fire, earth etc., and the parts and the premisses. The syllables, the seat, the medicine, the animal, the whole and the conclusion are forms. Having made this classification of forms and substrates, he quite reasonably adds that of these some are causes as the substrate, for example the 30 parts, while others are causes as the form. In this way he adds the efficient and the final.[226] And he presents the distinctions of the substrate, in so far as the way in which matter, strictly so-called, acts as substrate to what comes out of it, and what comes-to-be does so out of it, is different from the way in which the letters are substrate to the syllables and the parts to the whole, and different again from the way in which the hypotheses, which he calls premisses, are 320,1 substrate to the conclusion. Alexander says: 'Matter takes on the form by alteration, while the letters and the parts do so by composition (the syllable comes-to-be from the letters as from parts), while as for primary and simple bodies, which are called the elements of com- 5 pound bodies, viz. earth, water, air and fire, it is both by composition and by alteration that compound bodies are made from them. But all these are inherent and become causes for what comes from them; the premisses, however, are not inherent in the conclusion, but rather they are productive of it and are inherent in the whole syllogism and play the role of matter in it, while the conclusion plays that of form. 10 Perhaps in a sense the premisses are in the conclusion and it is all a single thing.'

This is one type of cause which, as has been shown, is manifold; another is the essence and the form. Such are the causes that he included with the material causes, and also the whole – not that which is with its parts, but that which supervenes on them. Compo- 15 sition can be of this sort. For this is something other than the constituents and supervenes on them as a form, as the syllable supervenes on the letters of the word and as the form due to composition supervenes on the elements of the body, for example a man or a fig-tree.[227] It is clear that the designation 'the form' fits all these. It is even more apt in the case of those things which have what is called underlying matter in the strict sense. We should remember that 20 Alexander says that the matter takes on the form by alteration. That is why the Peripatetics[228] say that the compound is made up of matter and form with these two basic components changing together in the process of the coming-to-be of the compound. Platonists say that matter is unchangeable, following what is said in the *Timaeus*, where

Plato says:[229] 'The same account must also be given of the nature that receives all bodies; it must always be called the same, for in no way does it depart from its character.' And it is clear that it does not alter. Perhaps, then, when Platonists talk about prime matter which is lacking all quality they do not admit that it can be altered. For that which receives a quality is altered by throwing off one quality and taking on another. But how could what is lacking in quality be altered? For this reason, then, they deny that the compound comes from matter and form, since these cannot change together, but they say that it is form in matter. Those from the Peripatetic school who understand proximate matter as the four elements or the seed and the menstrual fluid[230] quite reasonably say that this is altered, and they say that the compound of matter and form is a result of matter changing to form, and the form becoming materialised. 25 30 35

195a21 The seed, the doctor, the person who has given advice, and in general whatever acts, are all the primary source of change or its cessation. 321,1

Thirdly he again enumerates the efficient cause, giving examples here too, by means of which, I think, he demonstrates the distinctions within the efficient cause, as he did earlier with the material and formal. For the person who has given advice, the doctor and the seed are all efficient causes in different ways. The first of them initiates the process without putting his hand to the task; the doctor acts by doing the work himself; but the seed is, in a way, intermediate between the efficient and the material, since it acts by coming-to-be something else through a change to itself; the true agent, as Alexander agrees, must be distinguished from the product. We should not construe the word 'all' with the word 'acts', but should read it separately[231] and take it as referring to the three types of efficient cause. 5 10

195a23 Then there are those causes which are the end and good of other things; for the goal tends to be what is best for, and the end in view of, other things. It should make no difference whether we say this is good or apparent good.

These, then, are the causes, and they are of this number in species. 15

The fourth and last which he adds to the others is the final cause, saying 'the end and good of other things'. Alexander says that the reading 'Then there is the cause which is the end of other things, and also the good' makes good sense of the phrase 'of other things', which clearly refers to things that have an end. But perhaps the phrase 'of other things' is well taken as applying in common to 'the end' and to

20 'the good'. For just as a final cause such as health is the end in view for the other things that are employed for its sake, such as walking, the taking of medicine etc., so it is also the good of these things. To explain why the end and the goal are a good, Aristotle adds 'for the goal tends to be what is best for …'; for the goal is that which we aim for when doing everything else, and the aim is the good. What is more 25 to be aimed at, and what is more an end, is in fact more a good, 'what is best for other things'. Health is the end in view of exercise, happiness of health. Therefore this latter is the true end and what is best for other things.

He has said that the end and the goal are the good; he also said at 30 the start of the *Nicomachean Ethics*[232] 'Every art and every inquiry, and similarly every action and every pursuit, seems to aim at some good'; but that for the sake of which we do what we do is not truly and in every case good, but is in every case apparent good (by 'apparent' 322,1 I mean that which seems to be good, whether it is or not). That is why he added 'It should make no difference whether we say this is good or apparent good'. In this way the goal is said to be the good in the sense of the apparent good. For it is not good things themselves that move us (would that that were the case!), but the belief that they are 5 good. Therefore pleasure seems to be an aim and an end not because it is good (if it were it would not be damaging), but because it seems to be.

The causes being this many, Eudemus says:[233] 'The substrate and the agent of change are held by all to be the primary causes; many add the form and far fewer the goal.' But to us this seems to be the cause of most things, for it seems to be ubiquitous. Almost all our 10 actions are carried out for some goal and some[234] good. If on some occasion the doctor causes a fever, which is harmful in itself, at least he is doing it for the best when he is causing it. Similarly the person who chooses the lesser evil to avoid the greater achieves the good, like the person who chooses death in preference to other evils.[235] Such are 15 the ingenious arguments of some. But it is perhaps better to refute them, and this is easy.

195a27 The ways in which there are causes are many in number, but even these can be summed up under fewer headings. For the term 'cause' can be used in many ways, and even among causes of the same type some are prior and others posterior. For example, the cause of health can be said to be both the doctor and the professional; the cause of the octave is both the ratio 2:1 and number; there is always the larger genus in contrast to the particular species.

Furthermore, there is the cause *per accidens* and its genera;

for example, the cause of the statue is in one way Polyclitus, in another it is a sculptor, since being Polyclitus is incidental (*per accidens*) to being a sculptor; there is also the genus that is larger than the incidental, so that a man, or in even more general terms an animal, could be the cause of the statue. Of incidental causes some are nearer, others further; for example, a blond person and an artistic person could both be said to be the cause of the statue.

Earlier on when he said[236] 'In one way therefore "cause" means that from which anything comes-to-be as from something inherent' he meant by 'ways' the differences between the primary causes; but now 20 by 'ways' he means the differences within any one particular type of cause, for example the different senses of just 'matter', and of just 'form' and of the other causes – differences that belong within the one kind. He says that 'the term "cause" can be used in many ways even among causes of the same type', but that even these can be summed up as six differences for each type of cause. For with regard to each type of cause – for example material or formal – we say that the cause 25 is either *per se* or *per accidens*. Then each cause is taken to be either prior or posterior to another; for example, in the case of the efficient cause it is possible to posit prior and posterior causes of the same thing such as health, all of them *per se*, like the doctor and the professional; similarly in the case of the formal cause of the octave the ratio 2:1 and number; quite simply, any larger genus embracing 30 a particular species. Or rather the general and the particular are to be found everywhere; the more immediate being prior, the more remote being posterior. In each case the particular is the more immediate, since it is not general in the way that that which embraces it is. In the case of the material cause the particular lump of bronze 323,1 is the more immediate, while bronze in general or water or body is more remote; in the case of the final cause this particular instance of health is immediate, while health and state and quality are more remote and more general. Such a difference was indicated earlier 5 when he said 'also true on a less specific level'.[237]

So having considered these two species[238] – the individual and the general – in the case of what are said to be *per se* causes, he now offers the same species in the case of what are said to be causes *per accidens*, first explaining what causes *per accidens* are. The cause per se of the statue is the sculptor, and the cause *per accidens* is Polyclitus, 10 because being Polyclitus is incidental to being the sculptor. Therefore that which is incidental to the *per se* cause is a cause *per accidens*; the genus which pertains *per se* to what is incidental, in so far as it pertains to what is a cause *per accidens*, itself becomes a cause *per accidens*; for example Polyclitus is a cause of the statue *per accidens*,

15 although 'the man' pertains to Polyclitus *per se*. Therefore the man is a cause of the statue *per accidens*. Similarly with 'the animal'.

Having shown in this way, then, what *per accidens* causes are, he says that these too are divided between nearer and more remote, just as earlier he showed was the case with *per se* causes. For if Polyclitus 20 and the blond person and the artistic person are all causes *per accidens* of the statue, then Polyclitus is the more immediate cause while the other two are more remote, since being blond and artistic are accidents of Polyclitus.

I think we should note that in the case of the *per se* cause he took the more remote and the nearer as being what includes and what is 25 included in the sense of genus and species, while in the case of the cause *per accidens* he took it the same way, saying 'Furthermore there is the cause *per accidens* and its genera', although he gathered these under the heading '*per accidens*'. But in the case of the incidental and what pertains to the incidental he took the nearer and the more remote as something different. For the blond person and the artistic 30 person are the cause *per accidens* of the statue in a more remote sense, because being blond is an accident of Polyclitus, who is the nearer cause *per accidens*.

These, then, are the two pairs of species (one in the case of *per se* causes, the other in the case of causes *per accidens*); a little later on he adds a third pair while in the course of discussion. This is the one according to which *per se* and *per accidens* causes are spoken of either 35 as distinct from each other, as when I say on some occasions that the maker of the statue is the sculptor and on others that it is Polyclitus, or as combined with each other, as when I say that the cause of the 324,1 statue is the sculptor Polyclitus, not just Polyclitus or just the sculptor. That this third pair is added to the two other ones Aristotle himself makes clear a little later on when he puts the three in sequence.

5 **195b3** Besides all these senses, both those that refer to the cause proper and those that refer to it incidentally, there are also those that are spoken of as potential and those that are spoken of as actual; for example, the cause of building a house is either the builder or the builder actually building.

He adds a distinction common to all the causes he has listed (both those *per se* and those *per accidens*, as well as those that are distinct from each other and those that are combined) – the distinction between potential and actual; so the six original points of distinction 10 are now doubled by the distinction between potential and actual, and become twelve in all, which Aristotle himself clearly enumerates in the words: 'Even so, although these points of distinction are six in number, they can each be spoken of in two ways.' And to explain what they are he adds 'either individually' – clearly meaning *per se* – and 'or according to <its> genus', and again 'either *per accidens*' – clearly individually – and 'or as the genus of the incidental' and 'either as combined or as spoken of singly. And all of them <can be spoken of in two ways> either as actual or as potential.' The builder is potentially the efficient cause of the building of the house, even if he is not actually engaged in building it, but he is the actual cause when he is actually building it. 15

195b6 A similar account can be given also in the case of those things of which the causes are causes; for example this particular statue, or a statue, or in general an image; or this lump of bronze, or bronze, or matter in general. Similarly in the case of incidentals. And both the causes and their effects can be spoken of in combination, as when we say not 'Polyclitus' or 'the sculptor' but 'the sculptor Polyclitus'. 20

Even so, although these points of distinction are six in number, they can each be spoken of in two ways, either individually or according to genus, either *per accidens* or as the genus of the incidental, and either as combined or as spoken of singly. And all of them either as actual or as potential.

The continuation of the passage 'A similar account can be given also in the case of those things of which the causes are causes' refers to the effects; for those are the things of which the causes are causes. He says that the same distinctions hold good in their case as in the case of the causes. For some of these are more immediate effects, others more remote; some exist *per se*, others *per accidens*, for example, as an effect a statue e.g. of Milo[239] is more immediately this particular statue, less immediately simply a statue, and even less immediately an image. And the bronze (not taken as a material cause but as an effect of the miner) is in itself the immediate effect as this particular lump of bronze, less immediate as simply bronze, and least immediate as body or matter. He says this is also the case with the things that are incidental to the effects. For the things that are incidental to the effects are both incidental effects themselves, and of them some are more immediate effects, others less immediate. What are incidental to the more immediate *per se* effects, i.e. particulars, are more immediate incidental effects, while their species and genera are less immediate. For this particular russet, if the bronze which is produced has this colour, is a more immediate incidental effect, while russet and colour in general are less immediate. And effects, like causes, can be spoken of as not only distinct, but also as combined. For having listed the distinctions within causes and within 25 30 325,1 5 effects he adds the words 'And both the causes and their effects can be spoken of in combination', i.e. the causes and the effects.

10 Having listed the six points of distinction within the causes, I added, in the order I did, the same number of points of distinction within the effects in the order given by Aristotle in what he says subsequently. But he himself started with the distinction within the causes between *per se* and *per accidens* and between potential and actual; he then considered the same distinctions within the effects; and he finally applied the distinction between combination and 15 separation to causes and effects alike. Having thus enumerated the six points of distinction in the case of the causes and of the products, he subjects them all to the distinction between potential and actual. In combining it is possible to combine either *per se* with *per se*, or *per accidens* with *per accidens*, or *per se* with *per accidens*, and of these either immediate with immediate, or more remote with more remote, 20 or immediate with more remote. Yet the only combination which he himself seems to present is that of *per se* with *per accidens*, for example, if the phrase 'the sculptor Polyclitus' is used. For the *per se* cause, the sculptor, is combined with the *per accidens* cause, Polyclitus. Polyclitus is this sculptor, also mentioned by Galen,[240] who made the statue which had limbs perfectly proportioned both in themselves 25 and in relation to each other, so that for this reason this was called 'The Polyclitan rule'.

> **195b16** The distinction is this, that the actual agent, the individual, is being or is not being the agent in actuality at exactly the same time that the effect is being or is not being the effect in actuality; for example, the particular person applying treatment is applying it in actuality at the same time as the particular patient is receiving it, and the particular person building is building in actuality at the same time as the particular building is being built in actuality. But this is not always the case with the potential; for the builder and the building do not cease to exist at the same time.

Having said that the six points of distinction can be further distin-30 guished in two ways, by potentiality and actuality, he presents the distinction between potential and actual by saying that what are causes in actuality are, and are not, just that, simultaneously with their effects. Having said 'the actual agent' he adds 'the individual' not meaning that this is something other than the actual agent, but that the agent is the individual.[241] For this acts, affects and undergoes affection, while its species and genus do not. For none of the things which are generic in this way have substantial existence *per se*. But 35 what are individuals in actuality are also co-existent with each other, just as the particular builder is a builder in actuality at just the same time as what he is building, as long as he is building it. 'But this is not always the case with the potential.' He offers as clear evidence of this the fact that the builder and the building do not perish simultaneously.[242] For the building generally lasts longer than its builder. When someone is building there is always something being built, and when something is being built there is always someone building it; if no one is building, nothing is being built, and if nothing is being built, no one is building – so that they are what they are, or cease to be so, simultaneously. He does not here apply the term 'in potential' to what can be produced, for it is no longer possible for what has already come-to-be to come-to-be (such as the builder who has already built and the house which has already been built), but he uses the term to denote purely what is in contrast to the actual, namely what is not in actuality. For neither what can be in actuality nor what has ceased from being in actuality is in actuality. 326,1 5 10

> **195b21** We should always seek the highest cause of each thing (as in other questions). For example, a man builds in so far as he is a builder, but the builder builds according to his professional skill, which is, therefore, the prior cause. This is so in all cases.
>
> Furthermore, there are generic causes of generic things, and individual causes of individual things; for example, a sculptor is the cause of a statue, but this particular sculptor is the cause of this particular statue. Potentialities are the causes of potential things, and actualities of things that are being actualised.
>
> Let this be a sufficient discussion of how many causes there are, and in what way they are causes.

He says that the cause in the so-called strictest sense is the highest one; others call it the containing.[243] He makes this clear by his example. When we ask 'Why does a man build?' we get the answer 'Because he is a builder'. When we ask 'Why is he a builder?' we are told 'Because of his professional skill'. The enquiry of the cause ends there. That is why he calls it the highest, because we ascend to it and then stop. For in the strictest sense the cause of building is the professional skill of the builder. Just as in all other questions we understand what is strictly meant by a word, so too ought we to in the case of causes. For we understand by the word 'drachma' the genuine drachma, not a forgery, and by the word 'man' man in the strict sense, not a corpse. This is how we should also understand the term 'cause'. So the person who agreed with this would not suggest that nature was the strict cause of bodies. For even if it changes bodies, it is itself changed as it changes them. There is, then, some- 15 20 25 thing which changes it; if this is itself changed, we must look for what changes it, since everything that is changed is changed by something that changes it too, as Aristotle himself proved in book 7 of this 30 work.[244] But if it is itself unchanged, this would be the highest and strict cause of change. He adds that we should say that there are generic causes of generic things and that these correspond to each other,[245] and that there are individual causes of individual things, for example, a sculptor is the cause of a statue, and this sculptor is the cause of this statue. Similarly, the cause according to its potentiality corresponds to the effect in potentiality, and the cause according to the actuality to the effect according to the actuality. Alexander 35 comments on this passage that if the cause in potentiality were to 327,1 correspond to the effect in potentiality,[246] then knowledge would be simultaneous with its object, and perception with its object – these being relational things – at any rate if the term 'potential' were to be used in the case of that which can be produced, and not simply of that which is not in actuality; he meant this a little earlier on when he said that the builder in potentiality was the person who had stopped building and was no longer in the process of building. In this way the 5 problem in the *Categories* is resolved, the one concerning the shell at the bottom of the sea,[247] if the term 'in potentiality' were to be used in the case of 'that which can be produced'.

<CHAPTER 4>

195b31 It is said that luck and chance are among the causes, and that many things exist and come-to-be because of them. We must consider, then, in what way they are among these causes, whether luck is the same as chance or different from it, and in general just what each is.

Some wonder whether or not they actually exist; for they say that nothing comes-to-be because of luck, but that there is a definite cause for everything that we say comes-to-be as a result of luck or chance. For example, if you come into the market-place 'by luck' and find someone there whom you wanted to find but did not think you would, they say that the cause is your intention to go and shop. This is the same in the case of other so-called lucky occurrences; it is always possible to find some cause other than luck. In fact if there were such a thing as luck, it would really seem odd and a cause of perplexity why none of the thinkers of old, when discussing the causes of coming-to-be and passing away, made any definitive statement about luck; it appears that they too thought nothing was the result of luck.

He has stated what the nature and number of the causes are, and what senses and distinctions are observed in each case; since he now says 'It is said that luck and chance are among the causes, and that many things exist and come-to-be because of them', it is logical to ask whether these actually exist, what each of them is and into which category of cause as defined we must put them. <It is clear> that Aristotle did not idly say that they are said to exist. For obviously most people think that luck and chance are responsible for many things and even if the natural scientists say nothing about them, they do treat them and name them as if they did exist. When Empedocles[248] says 'For sometimes it chanced to be running in this way, often in another', and also 'However all things chanced to be', he seems to be giving luck a causal role. And those who claim that god or matter or good-and-evil are the basic principles will be forced to admit that their extension and their division in space and suchlike are the result of luck or chance. 10 15 20

When Democritus[249] too says 'The vortex was separated off from the whole, of all sorts of shapes' (how and by what cause he does not say) he seems to be generating it from luck and chance. And although Anaxagoras grants the existence of Mind, according to Eudemus,[250] he thinks that most things result from the chance. Some of the poets put almost everything down to luck, with the result that they make it the fellow of art, and say 'Art cleaves to luck, and luck to art'.[251] They also say that the man on whom luck smiles is wise. 25 30 328,1

In addition we can see that some of the things that are the products of art are also the products of luck. For health seems to result just as much from luck as from art; if the thirsty man drinks cool water he becomes healthy. But perhaps Democritus means that the cause is thirst, not luck. But most people put everything down to luck and pray to the gods for good luck. 5

So if luck and chance are said to be causes, then the person who thinks that he has given a complete account of causes must consider to which of the four stated causes he should ascribe luck and chance, and which of the various ways described of each of them is appropriate, and even before that whether luck and chance are identical or different, and if the latter, how they differ from each other. Even before that is the question of just what each of them is; for if the nature of each of them is not explained, any difference between them will not be clear. And even before asking the question what each of them is we should consider simply whether each exists;[252] for their being is not undisputed. And so, having set out the problems in order, and starting from the end problem found by the analysis, viz. whether luck and chance exist at all, he begins his demonstration from that. 10 15

First, he sets out the arguments on both sides, beginning with two on the part of those who do not think that luck and chance should be listed as causes, since they are not the causes of anything; the first

20 of these rests on proof, the second on opinion. The first goes like this: of all the things that come-to-be, including those that we say are a result of luck or chance, there are determined causes; neither luck nor chance could be causes of any of the things that have determined causes, since even if these are admitted to be causes, everybody agrees that they are indeterminate; therefore neither luck nor chance 25 could be the causes of anything that comes-to-be. It is possible to draw this conclusion in the first figure[253] according to the second of the undemonstrables.[254] It is also possible to draw the conclusion in the second figure: of all things that come-to-be the causes are determined; luck and chance are not the determined causes of anything; therefore neither luck nor chance are the causes of anything that comes-to-be. And he shows by means of the example that of all the things that 30 come-to-be, including those that we say are a result of luck or chance, there is a determined cause. For a man who goes into the market-place to shop, that is to spend time in the market-place or even to make his purchases, meets there someone in debt to him, whom he wants to meet but does not expect to, and gets his money. We all say that this happens by luck or as a result of chance; even so, the fact 35 that he goes wishing to shop is the efficient cause of what happens too; for it is because of that that he meets his debtor and gets his money. Similarly in other cases of what happens by luck it is always 329,1 possible to find some determined cause rather than luck. He did not think it worthwhile proving that even if luck is a cause it is indeterminate; for this is everyone's conception of luck and chance. The second argument would go like this: if luck and chance are the cause of what comes-to-be, there would be a paradoxical and perplexing 5 consequence – why on earth did the ancient thinkers, when discussing the causes of coming-to-be and passing away, not mention these? Anything that leads to a paradoxical conclusion is itself paradoxical, so that it would seem paradoxical for luck and chance to be the cause of any one of the things that come-to-be. But it is possible to draw the 10 following conclusion: if none of the ancient thinkers considered luck and chance to be causes, it is hardly likely that they are. But if the premiss is true, then so is the conclusion. That none of the ancients did think this he proves as follows: if those who discussed the causes of coming-to-be and passing away had thought that luck and chance were causes, they would have made definitions of them – but they did not do so.

15 Alexander criticises as ungrammatical the textual reading 'why none of the thinkers of old ... made no definitive statement about chance'. 'For he should', Alexander says, 'have written "made any definite statement", because the negative is already there in the word "none".' He obviously would have censured Plato even more strongly when he says: 'None of us can be of no value in nothing.'[255] We should

rather admire the passage for its clear and accurate expression, which commands respect. 20

196a11 But this too is surprising. For many things both come-to-be and exist as a result of luck and chance. Although everybody is perfectly well aware that it is possible to refer anything that comes-to-be back to some cause (as the old argument goes, which does away with luck), even so they all say that some things are the result of luck and that other things are not. Therefore they should have made mention of luck somehow or other. But they did not think that luck was any of the things like Love and Strife, or Mind, or Fire etc. It is equally odd, therefore, whether they did not think that it was anything at all, or whether they did but ignored it, especially since they did use it on occasions – for example Empedocles, who says that the air is not always separated out on top, but as luck dictates; for he says in his Cosmogony: 'Thus in its course it sometimes chanced to run like this, but often otherwise', and he also says that the parts of animals generally result from luck.

There are some who say that chance is the cause of both this world and all universes, claiming that the vortex and the movement, which separated out and established the cosmos in its present state, are the result of the chance. That really is surprising; for while they say that animals and plants neither exist nor come-to-be as the result of luck, but are caused by Nature or Mind or something else like that (for it is not any chance thing that comes from any particular seed, but an olive and a man each comes from the seed of its own kind), they say that the heavens and the most divine of natural phenomena result from luck and are not caused in the same way as animals and plants. Yet if this is the case, this very thing deserves close attention and some comment should be made on it. For what they say is odd in various respects, but it is even odder for them to claim this, when they can see nothing within the universe resulting from chance, whilst among the things that they say are not the result of chance they see much happening which clearly is. One would reasonably claim the reverse.

He makes a reply to the two arguments which show that luck and chance do not exist; he begins with the former, which does away with luck on the grounds that the causes of all things are determined while 25 luck is indeterminate. He opposes this by showing that such an argument achieves the opposite conclusion to that which it wants. For it wants to do away with luck, but without noticing what it is doing, it in fact confirms its existence. For if we are able in full

awareness to refer anything that comes-to-be back to determined causes, yet fail to do so in every case, even allowing that some things
30 are said to result from luck and chance, then it is clear that luck and chance are causal in some way other than a determined cause. For if we are never at a loss for a determined principle, yet in some cases ascribe causality not to the determined and evident principle but to luck, it is clear that we are ascribing causality to it as being something other than the determined principle. For why is it said that Achilles
35 did away with Hector as a result of choice, while Adrastus did away
330,1 with Croesus' son by luck?[256] Yet if the latter had not chosen to hurl his spear, Croesus' son would not have been killed. But he chose to hurl his spear, though not at Croesus' son but at the animal. Therefore we can say that choice was the cause of the hurling, but not of the killing. For in what way can he be said to have made a choice,
5 when he would never have despatched the missile if he had known what would happen? So it would be correct to say that such a death was the result of luck and not of foresight. Therefore in such happenings there is room for forbearance, since some of the things that are done by us are the result of luck, and some are the result of choice. For this reason the reality of luck ought to have received some
10 mention. In fact they clearly pay no attention to it, since they do not even name it, nor do any of the things named by them seem to be the same as luck – Love, Strife, Mind, Fire etc. (for these are determined causes), because Mind has a permanent predominance and the others dominate in turn.[257]

All this counters in a way the second argument too. The words 'as
15 the old argument goes, which does away with luck' seems to refer to Democritus, for even if he seems to have used luck in his cosmogony, in particular instances he denies that luck is the cause of anything, tracing things back to other causes; for example, he says that the cause of finding treasure is digging or olive-planting, or that the cause of the bald man's fractured skull is the fact that the eagle dropped a
20 tortoise so that the shell should smash. That is Eudemus' account.[258]

The second argument claims that the fact that the earlier natural scientists said nothing about luck is a strange and difficult consequence for those who do posit luck. In reply to this Aristotle says that the consequence is not strange for those who claim that luck is a
25 cause, but for the earlier natural scientists. For it is strange if they did not think that luck was a cause, which it so evidently is; for although they thought themselves wise and able to speculate about everything they failed to realise this fact, and although they did not think that luck was anything they still failed to persuade their fellow men that they were having recourse to an empty name. But if they did think that luck was something they omitted to explain what it
30 was and what power it had; it was indeed strange not to say anything

about this when they had declared their hand concerning causes. That they had some notion of lucky happenings is evident from their occasional use of the word. For example, Empedocles says that the air is not always separated out on top, but as luck dictates; for he says in his Cosmogony: 'Thus in its course it sometimes chanced to run like this, but often otherwise.' Elsewhere he says: 'Just as all things chanced to be', and he says that the parts of animals generally result from chance, as when he writes: 'The earth in particular chanced to be equally present in these' and 'Shining fire chanced to be part of short-lived earth', and 'Chancing on such an abundance of water by the devices of Love'. One could find much else in the same vein in the *Physics* of Empedocles to set alongside all this, such as: 'Thus by the will of chance did it think all things', and a little later on: 'According as they chanced to fall most sparsely.'[259] 35 331,1 5 10

Since Empedocles seems to have only a limited use for luck and does not explain what it is, he does not deserve so much consideration. But the followers of Democritus appear to be guilty of a double absurdity. First, although they say that chance is the cause of this world, they fail to tell us just what chance is. Why do I say 'this world', since according to them it is the cause of many or even an infinite number of universes? They say that 'the vortex and the movement, which separated out and established the cosmos in its present state, are the result of chance'.[260] Secondly, it is a cause of some surprise how it is that they maintain 'that animals and plants neither exist nor come-to-be as the result of luck, but are caused by Nature or Mind or something else like that'[261] – a cause which is determined and neither irregular nor irrational (this is obvious, since not just anything can come-to-be from just anything, but humans come-to-be from human seed and olives from olive seed; nobody says that a horse is sired by a horse as a result of luck, for what is regular and usually, if not always, the case is not the result of luck) – while in the case of the greatest and most divine of natural phenomena, viz. the heavens and the dance of the stars, in which there is nothing irregular and irrational, they trace the cause back to luck and chance, claiming that in this case there is no such cause as the one they posit for animals and plants, where it is determined and operates rationally and with regularity. 15 20 25 30

Even so, they say nothing about luck and chance. Even if they had, 'it would deserve their[262] close attention' if such things had to be ascribed to luck and chance, 'and some comment should be made on it'. For in addition to the absurdity of saying that the most divine phenomena in the universe are controlled by indeterminate causes, what they say next is even more absurd if one looks at it in a comparative light, since one sees that everything in the heavens is ordered and that nothing happens without determination or as a 332,1 5

result of luck and chance, while in the sublunary world – which they say is not the result of luck – many things happen as the result of
10 luck. According to their hypothesis the opposite should be true; the heavens would be moved irregularly, while the nature of the world of coming-to-be would be entirely ordered; and heavenly bodies would be shorter-lived (such is the nature of the products of luck), while things on earth would be longer-lived. But we can see that heavenly bodies surpass from earthly things much more in order and duration
15 than in size.

Eudemus defends those who do not list luck among the causes, saying:[263] 'It was quite rational for them to omit luck, since it is *per accidens*. Such things do not even appear to exist; only what is *per se* exists.' In all these passages Aristotle's reporting is clear and orderly. Alexander's commentary on them seems to me to be rather inade-
20 quate and incoherent, perhaps because the copyist has confused the order of what he said. But I consider that those who criticise Aristotle on the grounds that he does not mention the efficient cause of the heavens should listen carefully to what is said in this passage. For if he censures those who say that the heavens and the most divine of
25 natural phenomena result from luck and are not caused in the same way as animals and plants, then it is clear that he wants these things to come-to-be, even if not in a temporal way, then at least as the result of some efficient cause which is entirely determined and ordered, since he censures those who say that they come-to-be as the result of chance, and since everything that comes-to-be does so because of
30 some agent. Such an agent would be intellect or even something above intellect, since he concedes that either nature or intellect or something of the sort is the cause of animals and plants.

196b5 There are those who think that luck is a cause, but that it is unclear to human thought, since it is something divine and rather numinous. Therefore we must examine luck and chance to see (a) what each of them is, (b) whether or not they are identical, and (c) how they fall within the scope of the causes we have defined.

35 Some people, even though they do not say outright that luck exists,
333,1 are nevertheless forced by their own statements to admit that it has substantial existence. Others agree that luck obviously does exist, and say that it is a cause. But they cannot say just what it is, since they think that it is unclear to human thought; it is something divine
5 and numinous, which is why it is beyond human thought, as the Stoics apparently claim. That this is a common opinion is clear from the fact that many people worship Luck[264] as if it were a god, build temples to it and sing hymns to it. The notion of Luck as something divine

seems to have been current in Greece even before Aristotle and not to have been first conceived by the Stoics, as some think. In fact in the *Laws*[265] Plato says: 'God rules everything, and together with Luck 10 and Circumstance controls all human affairs.' For certain cities to honour Luck and build temples to it seems to have been a later fashion, since earlier writers do not record instances of temples or festivals of Luck in any cities. Even so, we know that the name 'Luck' was revered in olden times; at Delphi the pre-amble to the inquiry 15 was 'O Luck and Apollo, will you answer this request?'. Luck is also mentioned in Orphic pronouncements.[266] Therefore if some made luck the cause of the more divine bodies in the universe without claiming that it really was something, if others named it without giving any account of it, and if yet others assumed that it was something divine, 20 then all this makes it clear that it is something. It is worth enquiring just what luck and chance are, whether they are different from each other, and how they fall within the scope of the causes that have been determined – i.e. whether they are material or formal rather than efficient or final causes, and whether they belong to the category of *per se* or *per accidens*. 25

<CHAPTER 5>

196b10 First, then, since we see that of things that come-to-be some do so always in the same way and others usually, it is obvious that neither luck nor the result of luck can be said to be the cause in either case – either of that which is of necessity and always, or of that which is usually so. But since there are things other than these which come-to-be, and since everybody says that these are the result of luck, it is obvious that luck and chance are something. For we know that such things are the result of luck, and the results of luck are of this sort.

[This section of Simplicius' commentary runs on as far as 197a8, although the *lemma* in the Diels text runs only as far as 196b17. In fact most of the passage from 196b17-197a8 is quoted by Simplicius in the body of the commentary.]

He has proposed to speak about luck and chance, and has undertaken first to enquire whether or not it exists; he has shown that it does 30 exist, basing his argument on the common testimony not only of all those who say that it is a divine and numinous cause and who speak of luck as a cause (even if they fail to articulate their thoughts precisely), but also of all those who attempted to deny it – for even they were seen rather to confirm its existence. He then moved on to the ensuing questions, putting forward for discussion what each of

35 them is, how they differ from each other and how they fall within the scope of the causes he has determined.

334,1 He presses his first proof by again showing that luck and chance are something, not only on the basis of common testimony, but also on the basis of the facts themselves (in concluding this argument he says that it is clear that luck and chance are something); and at the same time he makes a start with his argument about their existence.
5 I think his reasoning was as follows: he takes it as agreed that anything that comes-to-be is always caused, and that its cause is what brings it into being; he then shows that things that are not generally the case both exist in the scheme of things and come-to-be as a result of luck and chance: if, then, things that are not generally the case exist in the scheme of things and are the result of luck and chance,
10 then it is obvious that luck and chance are something. By making a division of things that come-to-be, he shows that things that are not generally the case do exist, since things that are not generally the case could not belong to the category of the ungenerated. For of things that come-to-be some do so always in the same way, like the rising and the setting of the sun, while others do not always do so. And of
15 these some usually do so, such as natural events in the sublunary world; for example, man begets man, horse begets horse, and the man who goes into the market-place for a specific purpose finds what he is looking for; some things do not occur usually, but either half or less of the time. What occurs half of the time occurs less often than what occurs always or usually. That is why Aristotle, I think, continues to make no distinction between what occurs half of the time and what
20 occurs less than this, in contrast to what usually occurs. That what is in contrast to the usual happens as a result of luck and chance is clear from the fact that everybody both says and knows that such things are the result of luck – and not only is this the case, but also the converse, that the results of luck are like this. For luck does not belong to the realm of the ungenerated eternal real beings – for
25 nobody says that Mind or any of the unchanged causes are the result of luck – nor does it belong to the realm of what comes-to-be but always comes-to-be in the same way, nor does it belong to the realm of what usually comes-to-be, but to that of what occurs less than usually. Having shown that there are such occurrences, he concludes that luck is something and that it belongs to their realm. Luck is to be found where the products of both art and nature display deficiencies; for example, in the case of medicine, luck slips in where skill
30 fails to reach. Therefore the results of luck count for much with poor craftsmen, but for very little with excellent craftsmen and with the most precise of crafts. It does not figure at all in the foremost and noblest of the things that exist according to nature (for they always abide unchangingly by their nature), but it does figure in the realm

of coming-to-be and passing away, in so far as nature does fail, where it does so.

It is clear that he is not yet making any distinction between luck 35 and chance – even if he only names the former – but is discussing both. Furthermore, even if he seemed in that passage, where he said 335,1 'everybody says that these are the result of luck', to be have been constructing his argument from the testimony of the majority, he is now arguing as if the case were evident. That is why he adds: 'Of things that come-to-be, some do so for the sake of something, others do not.' From the fact that there are things that occur less of the time 5 and that luck and chance are to do with these he showed that these things do exist; he then proceeds to explain further the way in which luck and chance are the causes of these things, that just as something may happen *per accidens* alongside what happens in accordance with choice and nature, so luck and chance are causes in just this sense; from this he shows just what they are. For in response to the question 'Do they exist?' a sufficient reply lies in the existence of their effects; 10 but to answer the question 'What are these causes?' one needs to know the manner of the existence of the effects.

He further divides all things that come-to-be into what is for the sake of something and what is not; for there is no purpose behind many things that happen (it is through bad habit, for example, that people pluck their hair, bite their nails even when they are short, and pick their noses) – when they have no point of reference when they 15 happen. When he asks what things are for the sake of something, he says that they are those things which are according to choice (including the products of art) and those which are according to nature, for both choice and nature act for the sake of something, as he will prove at length. He is not using choice in the sense that he does in the *Topics* and *Nicomachean Ethics*.[267] He aptly says that the term 'for the sake 20 of something' includes both what might have been by intention and what might have occurred naturally, even if it was not done so but could have been (for example, going into the market-place could have been for the sake of recovering what is owed to you; and even if this was not the intention, the recovery of the debt is said to be one of the things that are for the sake of something). Since 'things for the sake 25 of which' are like this, then both luck and chance belong to that category. If, then, it has been proved above that luck and chance belong to the category of things that happen less of the time, the conclusion in the third figure[268] is that some of the things that happen less of the time happen for the sake of something, as Aristotle himself points out in the words 'Therefore it is clear that among what is neither of necessity nor usually the case there are some things to which the term "for the sake of something" can be applied'. The words 30 'Therefore it is clear' denote a conclusion, in which he deduces that we say that such consequences are the result of luck, whenever they occur *per accidens*, in the case of things that are for the sake of something less of the time.

336,1 That he means by the phrase 'that which is for the sake of something' not always what we actually do for the sake of something, but also what can be done for the sake of something, he makes clear a little lower when he says: 'Let this much now be clear, that both belong to the category of what is for the sake of something. For example, he would have come to recover his debt when his debtor was making a collection, if he had known this; but although he did not 5 come for the sake of this, he did come and do what was necessary to recover the debt *per accidens*.' Alexander, however, says: 'This was not the result of luck purely and simply, but is an example of good luck.' The words 'both belong to the category of what is for the sake of something', when used earlier, did not apply to what is in accordance with choice (as some suppose), but to luck and chance,[269] as is 10 made clear in what is expressed in the same words in the subsequent argument, where he says: 'Let this much now be clear, that both belong to the category of what is for the sake of something'– namely the two things that the argument is about, luck and chance; he had promised that he would distinguish between them later on. If anyone imagined that these words did apply to what is in accordance with choice and not as I have suggested, then the subsequent conclusion 15 did not follow. For the conclusion as stated does not follow from the proposition that what results from luck or is according to nature is for the sake of something, since the argument will not hang together. But he quite rightly thought it worthwhile showing that 'for the sake of something' applies to certain of the things that occur less of the time, because such occurrences might have seemed to some to be intentionless; without this proof it did not follow, in the case of what happens less of the time, that the results of luck were in the category of what is for the sake of something.

20 Eudemus demonstrated that luck is in the category of what is for the sake of something as follows: 'Luck occurs where there is good or bad fortune;[270] these are cases where one achieves or fails to achieve some good, and where "for the sake of something" applies, since anyone who acts for the sake of something desires some good.' This argument put only luck, and not chance, in the category of what is for the sake of something. But when what happens for the sake of 25 something, i.e. the results of intention and nature, becomes the cause of something *per accidens* in the case of what happens less of the time, then we say that what occurs does so as a result of luck and chance. The phrase 'that for the sake of which' means not that the visit *was* made for the sake of something such as shopping, as Porphyry says, but that it *would have been made* for this purpose. For Aristotle says

'he would have come when his debtor was making a collection, if he had known this'. Things are caused *per accidens* as a result of thought and nature when not the intended result but something else occurs; for example, if I go out intending to do some shopping but meet my debtor and recover my debt, or if the stone drops in order to find its proper unity[271] but on its descent strikes someone on the head, or falls in such a way as to form a seat. 30 35

He has said that whenever what happens as a result of intention and nature (such as the visit to the market or the fall of the stone) becomes the cause of something *per accidens*, then what happens as a result of them is said to happen as a result of luck and chance, since it did not happen in keeping with the essential description.[272] He quite reasonably goes on to show that in general some causes are *per accidens*, as he said earlier, and he now reminds us of that using the example of the house. The *per se* cause of the house is the building skill and the craftsman who exercises it, while the *per accidens* cause is the fair-skinned or the artistic man. Alexander says: 'Aristotle says that just as anything that exists is one thing *per se* and another *per accidens* (by "being what it is *per se*" he means the substance, and by "what it is *per accidens*" he means the attributes of the substance), so a cause is one thing *per se* and another *per accidens*.' Perhaps, however, we should understand 'anything that exists' in a simpler sense as 'anything which is caused', so that he would be meaning that just as what is caused is one thing *per se* and another *per accidens*, so too is the cause. For the categories other than substance are *per accidens* in that they always subsist in a substance, but are not always present *per accidens*.[273] Luck differs from the other *per accidens* causes in that each of them is said to be a cause *per accidens* in that they are incidental to the *per se* cause[274] (as fair-skinned is the cause of the house *per accidens*), since being fair-skinned is incidental to the builder, who is the *per se* cause of the house. But in the case of luck one and the same action is the *per se* cause of one thing and the *per accidens* cause of another. Digging is the *per se* cause of planting but is the *per accidens* cause of finding the treasure. So in the one case there is a distinction within the causes (one is a cause *per se* and the other *per accidens*), while in the other the cause is one and the same, but the results are different, one following *per se*, the other *per accidens*. But even if the cause seems to be the same, as in the case of digging, perhaps it is still a cause in different ways – of the planting *per se* and of the discovery of the treasure *per accidens*. In this case it is not only the efficient cause that is meant (as it was in other cases) as *per accidens*, but also the end in view. For just as digging with the intention of planting is, if the occasion arises, the *per accidens* cause of the discovery of treasure, so the discovery is the *per accidens* result of digging with the intention of planting. A *per accidens* result is said 337,1 5 10 15 20 25 30 to be the outcome of luck, while what is incidental to the agent is luck.[275]

Yet luck embraces the indeterminate no less – if not more – than the other *per accidens* causes. For the discovery of treasure could supervene on a large, even a limitless number of causes that are there
35 for the sake of something quite different. One might dig in order to plant, to lay foundations, to find water, to construct a conduit, a pit
338,1 or a grave, or for countless other reasons. So if the result is not indeterminate but prescribed, then it is no longer the outcome of luck, nor is luck the agent.

Even though Democritus usually takes a different standpoint, he seems to concur with general conceptions of luck. For he says that he
5 did not find the treasure as a result of luck, but because he was digging in order to plant, to find water or some such thing, in the course of which he would have found it. Each of these activities would be and become the cause in the same way. Therefore since there are many activities which, according to Democritus, could become causes, the cause would be indeterminate in this way too, and not *per se*, but *per accidens*, since it is incidental to other things for the sake of which
10 the digging was done. If, then, both the outcome of luck and luck itself are indeterminate, then it is clear that it belongs neither to the realm of ungenerated and eternally unchanged beings which do not admit of any movement or alteration, nor to the realm of the things that come-to-be and which do so of necessity; nor does it even belong to the category of what is the case most of the time (for this too is more determinate), but to the category of what is the case less of the time.
15 If it is the consequence of some preceding change, and if all change belongs to what is the case most of the time, and if chance is the result of either intention or nature, and if each of these is for the sake of something, then it is entirely reasonable for luck to be placed among the things that occur for the sake of something, not those that occur *per se*, but *per accidens*, since it is indeterminate. For just as *per se* causes are determinate, so *per accidens* causes are indeterminate.
20 The builder is the cause of the house in a determined way, but the fair-skinned and the artistic are nevertheless among the *per accidens* causes.[276] So if one were to think that the *per se* causes were the only ones, nothing would occur as the result of luck; but if *per accidens* causes too are admitted, then things could happen as a result of luck.

He shows by means of the example just what this means. We put
25 down to luck the consequence of an action which could have been taken to bring about that consequence, but which on that particular occasion was not taken for that particular purpose. The man could have gone out in order to recover the debt. If he had gone out for that purpose and recovered it, he would not have recovered it by luck; but he would claim this when he goes out for some other purpose and then

finds his debtor making a collection and so recovers his debt. This is an example not simply of luck, but of good luck. For when an action taken by choice results in an outcome for which the action might have been taken as a primary objective, it is clear that the outcome is something one would have both desired and chosen. But sometimes something bad and undesirable results, when the people were not acting or would not have acted to achieve it; this is no less the result of luck than the other case, and luck is the cause, albeit bad luck, since the outcome is what you would avoid and not choose to have happen. 30 35

Alexander says: 'A general definition of luck would be "a result that follows an action that is taken by choice and for a purpose, but which does not fulfill that purpose for the sake of which the action was taken by choice as a primary objective".' But this is not luck, but the result of luck. Luck, as Aristotle himself defines it, is a *per accidens* cause in the case of things that are subject to choice. It is clear that luck is an efficient cause, as will be stated, but efficient *per accidens*, just as the outcome that results from luck is *per accidens*. He says that that which becomes the *per accidens* cause of a lucky outcome – for example going into the market-place – cannot belong to the category of the necessary or of what occurs most of the time. For when someone makes a daily visit to the market-place out of necessity or most of the time, the fact that his debtor meets him on some occasion is not the result of luck. For this was bound to happen at some time, and was quite probable. But if he visits the market-place only rarely or just once and for some other purpose, and then his debtor meets him, then this is the result of luck. Aristotle quite rightly added this to the exposition of the definition, since luck belongs to what occurs less of the time. 339,1 5 10

The outcome which is the result of luck, he says, such as the recovery of the money, does not depend on inherent causes, but does depend on what would be chosen, and is the outcome of intention. For it does not depend on such inherent final causes as do the ends of things that come-to-be by nature, for the sake of which the preceding stages take place in the actual thing which is coming-to-be by nature; rather they belong to the category of what supervenes from outside. That is the character of what happens as a result of luck and of choice. 15

Alexander is aware of an alternative reading: '... is something that was not chosen and was not the result of intention.' He says: 'Even if this were what Aristotle wrote, he would still say that it is the result of luck whenever something occurs when we have not made a choice about it, have not formed an intention about it and when our prior actions have not been for the sake of it. For a little earlier he has said that what happens as a result of luck can lie within the scope of choice and intention.' He adds the words 'but if he had made a choice for the 20 25

sake of this', which seems to suit this reading. For the words 'but if he had made a choice' seem to have been contrasted to 'would not be chosen'. The former reading distinguishes the outcome of luck from the natural; for natural outcomes are not chosen, even from outside,
30 but are inherent in the actual things that have a nature. For in the case of things that occur by nature it is not a goal aimed at but some other outcome that results from natural changes happening for the sake of something else, such as, in the case of a man, being six-fingered or ox-faced.[277] But such outcomes are not the result of luck.[278] But we say that a lucky outcome occurs where choice might be involved, and that it occurs from outside in that it is not inherent in
35 us. For the recovery of the debt is not by nature inherent in the person making the recovery, since it is not inherent in him by nature to make the choice and go to that particular place. Even if the man were by nature such as to make the choice, this particular choice would not
340,1 be natural. What Aristotle says can be understood in this sense: the outcome – the recovery of the debt – is not an outcome or result of inherent causes, but nature is an inherent cause, for nature is a source of change in those things in which it is present primarily; therefore that which happens as a result of luck does not have nature as its efficient cause, but intention and choice. Therefore intention
5 and luck are to do with the same thing; luck is to do with choice and action, as will be shown, while choice is always accompanied by intention.

> **197a8** Therefore the causes from which the outcome of luck can occur are of necessity indeterminate. Consequently, luck seems to be the cause of what is indeterminate, and to be unclear to man.

10 He has said just what luck is, and has explained it by means of a definition which he has arrived at on the basis of proven facts, that luck is a cause, taking it in a general sense, and that it is 'a cause *per accidens* in the case of what happens by choice for some end' – although if it is *per accidens* it is also less often the case.[279] Having said this, then, he wants, as it seems to me, in his usual manner to show next that anyone with anything to say about luck is in agree-
15 ment with him. He does this by showing that the notion of luck which he has given is in harmony with what is said about it. For in fact luck is said by some to be a cause, but only a cause of the indeterminate; others claim that it is not in any way a cause of anything that occurs; others say that it is, but that it remains 'unclear to human thought'.
20 Luck is said by some to lie outside calculation. 'All these views', he says, 'are correctly stated', and they fit in with what I myself have said. For since luck is a cause *per accidens*, and since *per accidens*

causes are indeterminate, they differ in this respect from *per se* causes, in that in the case of the latter it is possible, when giving a definition, to say that one particular thing is the cause of a particular effect, while one cannot do so in the case of *per accidens* causes. Take the case of the person getting his money back; granted he came <to the market-place> – but not for that specific purpose, which we say is the outcome of luck – then its causes can be without limit; for he could have come to meet someone, or to take part in a lawsuit, or to see some spectacle. There could be a countless number of such causes for his coming to that place, and all these will be *per accidens* causes for his getting his money back – no one of them is more a particular cause than any other. For the only *per se* cause of his going to the market-place and getting his money back would be going there in order to get it back. 'Perhaps', says Alexander, 'it would for that reason be more in keeping with what he has already stated to say that what happens for the sake of something is the *per accidens* cause of that for the sake of which it does not happen; for that for the sake of which it does happen is something single, while the things for the sake of which it does not happen can occur and be contingent to it in countless numbers, one of which is getting his money back.' He says it is more in keeping because it has been stated that luck is the cause of the indeterminate. Clearly such a notion understands the cause as an end, i.e. the outcome of luck, to be something indeterminate. But Aristotle says that it is not the final causes of lucky outcomes that are indeterminate, but the efficient causes, since it is from these that the outcomes of luck might result.[280] Furthermore he proceeds to adduce examples, saying that the *per accidens* causes for his going to the market-place and getting his money back are infinite in number. Yet if, as he says, 'the causes from which the outcome of luck can occur are indeterminate', then luck could quite reasonably be said to be indeterminate, since it is interpreted as being among the indeterminate causes, and it is and is said to be the cause of an indeterminate result. But if both the outcome of luck and luck itself are indeterminate, then it is reasonable to say that it is 'unclear to human thought'. For the indeterminate cannot be grasped by reasoning. It is worth noting that he first says that luck seems to some to be 'unclear to human thought' not because it is indeterminate but because it is 'something divine and numinous'. Perhaps those other thinkers had some inkling of this in that they were not able to produce any clearly defined cause of what happens in this way.

197a10 But in a way it might seem that nothing is the outcome of luck. For all these views have been correctly stated, since they have been put with good reason. In one way things can be the outcome of luck; they can happen *per accidens*, and luck is a *per accidens* cause. But it is not the cause of anything pure and simple. For example, the cause of a house is a man *qua* builder; *qua* oboist he is its cause *per accidens*. In the case of the man going and getting his money back, if he did not go for this purpose the causes are indefinite in number, such as wanting to meet someone, to take part in a lawsuit or to see some spectacle.

20 He adds this third point, that even those who say that nothing is the outcome of luck would in a way be speaking correctly. For if anyone were to claim that the only causes are those which are spoken of in the strict sense and *per se*, and not *per accidens* causes as well, then luck will not seem to be the cause of anything, since it is one of the so-called *per accidens* causes. Therefore he says that all these views 25 are correctly spoken – that it is unclear to man, that it is the cause of nothing, and that it is the cause of what is indefinite – for they all embrace the plausible account, and are all in harmony with the given definition of luck. Clearly even in itself this universal concord is positive proof that there is such a thing as luck. He proceeds to adduce an example to show that luck is a *per accidens* but in no way a *per se* 30 cause, and that it is an indeterminate cause, since 'in the case of the man going and getting his money back, if he did not go for this purpose, the causes can be indefinite in number'.

Having said that luck is a cause *per accidens*, and never a cause 342,1 pure and simple, he adduces an example of a *per se* and a *per accidens* cause, saying that 'For example, the cause of a house is a man *qua* builder' (clearly meaning *per se* and in the strict sense), but '*qua* oboist he is its cause *per accidens*'. He then adds the example of the *per accidens* outcome of luck, the man going and getting his money back 5 but not going for this purpose. It is worth noting that not everything which is *per accidens* is a cause of luck, even if choice is involved, and if it is less often the case, and if one is acting for the sake of something, just as not everything which is *per accidens* is the result of luck. The outcome of luck is that which occurs as the result of a *per se* cause involving choice but which does not occur *per se*; but luck is the *per* 10 *se* cause in itself even if it is not taken as being *per se*. For to choose to go out in order to greet a friend, becoming the *per accidens* cause of getting the money back, becomes a piece of luck, and the retrieval of the money is the outcome of luck.[281] But that which is contingent upon the *per se* cause is something other than it (as being a oboist is contingent upon being a builder), and if it is not a piece of luck, then nor can the outcome of it be said to be the outcome of luck. For the 15 builder had no other intention or aim, whereas that would need to be seen in those things that are the outcome of luck.

197a18 It is correct to say that luck is something that cannot be calculated. For calculation concerns that which is always or usually the case, while luck is to do with the converse. Since their causes are indeterminate in the way explained, so too is luck.

He says that it is correct to say that 'luck is something that cannot be calculated'. For calculation concerns the determined, so that the 20 indeterminate cannot be calculated. If, then, luck is indeterminate, and if the indeterminate cannot be calculated, then it is reasonable to say that luck cannot be calculated. He shows us that luck is indeterminate by reminding us that luck belongs to what is less often the case. For if luck belongs to causes which are less often the case, and if what is less often the case is indeterminate, then luck is indeterminate.

197a21 Even so, in some instances we could ask ourselves the 25 question: Could anything whatever be the cause of a piece of luck? For example, could the causes of health be fresh air or sunshine, and not a haircut? For of *per accidens* causes some are more proximate than others.

He has said that luck is an indeterminate cause. He now says that it is problematic whether anything that chances[282] might equally be the cause of luck, i.e. the cause of the outcome of luck, or whether some things might be causes more so than others, as is the case with the other *per accidens* causes. For Polyclitus is a more immediate cause 30 of the statue than the man is, since in the case of *per se* causes too the sculptor is a more immediate cause than the craftsman. When someone is ill and has his hair cut,[283] and because of the loss of his 343,1 hair either receives some wholesome fresh air or some sunshine (i.e. some warmth from the sun, since the rays of the sun cause warmth, hence our term 'sunbathing') and regains his health, is the fresh air or the sunshine equally the cause with the hair-cut of his happening to become well again, or is the fresh air or the sunshine more the cause? It is not the hair-cut, since of *per accidens* causes too some are 5 more proximate than others. For if this is so, the outcome of luck is not indeterminate purely and simply, nor is luck purely and simply an indeterminate cause, since the more proximate cause is determined. The problem, then, is of this nature, I think, leaving us to find the solution which too would be of this kind: when we take and compare the *per accidens* causes, then <we see that> things that 10 happen to occur are not all equally the causes of the outcome of luck, but that the more proximate ones are more the causes. But when we ask the question absolutely, 'What could the cause of the outcome of luck be?', then it is impossible to define it and say that it is such-and-such, since it is possible to find causes that are infinite in number and indeterminate.

Alexander puts the question: If luck is a *per accidens* cause, and if 15 that which is contingent upon the *per se* cause is a *per accidens* cause, will there be some *per se* cause of an outcome of luck, so that what is contingent on this is luck? He says that coming is the *per se* cause of the purpose of coming, but the *per accidens* cause of what occurred, which is the outcome of luck. Then the coming is, as a cause, a piece of luck – when it is taken not *per se* pure and simple, but as a *per* 20 *accidens* cause.

> **197a25** Luck is said to be good when something good results bad when something bad results; but it is called a godsend or a disaster when the results are of some significance. So narrowly to miss a great evil and narrowly to miss a great benefit is to enjoy a godsend or to suffer a disaster, since the mind proclaims them as already yours; for what is so near seems to be no distance off at all.
>
> Yet it would be fair to say that a godsend is fickle; for luck is fickle, since no outcome of luck can be what is always or usually the case.

The difference between good and bad luck on the one hand, and a godsend and a disaster on the other, is clear. For universally the 25 prefix god-[284] comports significance and excellence; it is the wholly, not the moderately, celebrated man, that we call godlike, and similarly with the man of god-given intelligence. In the same way the prefix dis- indicates a corresponding significance. Aristotle shows that a godsend is significant by reckoning it, I think, in what is effectively a syllogism as follows:[285] A godsend or a disaster, even if it achieves the greatest benefit or the greatest evil by a hairsbreadth, is still a godsend or a disaster; what achieves by a hairsbreadth is no 30 less what it is proclaimed to be; these are significant cases because in significant matters to achieve the whole by a hairsbreadth seems to be as good as achieving it. It is in this way, then, that a godsend and a disaster are spoken of as being significant.

Porphyry and Alexander according to a first interpretation took it 344,1 more in this sense, the term 'by a hairsbreadth' in this passage being understood not in the sense of coming close but being prevented, but in the sense of achieving the whole good or the whole evil by a hairsbreadth, since we still call it a godsend even if someone achieves the greatest good by a hairsbreadth, just like someone getting the whole amount; and we call it a disaster even if he narrowly fails to 5 avoid the greatest evil, since the hairsbreadth makes no difference.

It is possible to interpret the passage in a different way.[286] If the term 'by a hairsbreadth' is usually used in cases of significance, and if we call unlucky those who, with a hairsbreadth's difference, might have achieved some great evil, and lucky those people who, with a hairsbreadth's difference, might have achieved[287] some great good, then it is clear that the enjoyment of a godsend or a disaster depends on the significance. The Spartans honoured as saviours the so-called 10 'hairsbreadth gods',[288] because they saved men who were about to risk their lives by allowing them to come within a hairsbreadth of risking them. Alexander tends more to this interpretation.

The addition 'since the mind proclaims them as already yours; for what is so near seems to be no distance off at all' seems to fit in better with the former interpretation. For how could the mind proclaim as being 'already yours' something which might, by a hairsbreadth's 15 difference, have happened? If it were to proclaim this, then the person coming within a hairsbreadth of a great evil would suffer disaster, since the narrowness of the gap would amount to nothing of any significance. Yet we actually say that he enjoys a godsend. Similarly the person coming within a hairsbreadth of a great benefit would enjoy a godsend. Yet we actually say that he suffers disaster, because he has been deprived of a great benefit – and that is a great evil.

In general such significance, where such a hairsbreadth difference 20 is seen, is not called significant because it itself is significant, but because the significance or lack of it are judged by the outcome. Perhaps, then, he does not say that to enjoy a godsend or to suffer a disaster is a matter of scale because of the hairsbreadth difference, but because even those who come within a hairsbreadth fall into a great evil or a great good, since we do not call them anything[289] because they have come within a hairsbreadth; but because they have come close to a great evil or a great good and have not achieved it we 25 call some lucky, others unlucky – like those saved from death. He makes it clear that this is his meaning when he says 'So narrowly to miss a great evil and narrowly to miss a great benefit is to enjoy a godsend or to suffer a disaster, since the mind proclaims them as already yours'; and he relates enjoying a godsend to coming within a hairsbreadth of a great evil, and suffering a disaster to coming within 30 a hairsbreadth of a great benefit. He has previously shown this from the case of a person who already has good or bad luck, and he now demonstrates the same point from the case of those who have come within a hairsbreadth of losing them.

He says that the fact that good luck is said to be fickle is in keeping with the notions expressed about luck. For if luck is fickle, then it is clear that a godsend too is fickle. It is clear that luck is fickle from 35 the fact that the outcome of luck does not belong to the realm of what 345,1 is always or usually the case, but to the realm of what is less often the case. He presents all that has been said about luck because the notion of luck current in people's thinking was the one articulated by the argument offered by Aristotle.

5 **197a32** Therefore, as has been stated, both these – luck and chance – are *per accidens* causes operating in the case of things that have the possibility of coming-to-be, but not always and not for the most part, and where things happen for the sake of something.

After the discussion of luck, it followed that he should explain further about chance. What is it, and how does it differ from luck? First he presents the common ground, that both are *per accidens* causes 10 operating in the case of things that happen less often and as possibilities, but not always,[290] i.e. of necessity, and not in the case of things that come-to-be for the most part; and where things happen less often, in the case of what comes-to-be for the sake of something. Since what is possible[291] is predicated of what is necessary and what is for the most part (for it has the same meaning in both cases), he said that chance operates within the area of what is possible; he defined in 15 what sense it is possible, and added 'not only[292] always and not for the most part' in order that we should understand that by the term 'what is possible' he now means 'what is less often the case'.

<CHAPTER 6>

197a36 They differ in that chance has broader connotations; for every outcome of luck is the result of chance, while the reverse is not the case. For luck and the outcome of luck are to do with such things as enjoy good luck and in general with rational activity. Therefore luck must be to do with matters concerning such activity. Evidence of this is the fact that good luck is thought to be the same, or nearly the same, as happiness, and happiness is a sort of rational activity, for it is successful activity. Consequently, whatever cannot act in this way cannot achieve anything as the result of luck. Therefore nothing inanimate, nor even a beast or a child, can achieve anything as the result of luck, since they do not exercise choice. They do not even enjoy good or bad luck – except by similarity; Protarchus says that the stones that form an altar enjoy good chance because they are revered, while their fellows are trodden underfoot. Even these things can, in a way, undergo something as the result of luck, when the agent achieves something in their case as a result of luck. Otherwise this cannot be the case.

Having said what luck and chance have in common, he now proceeds to differentiate between them, saying that 'chance has broader con- 20 notations; for every outcome of luck is the result of chance, while the reverse is not the case'. For luck belongs solely to matters concerning rational activity,[293] i.e. where choice is involved, as he has explained in the *Nicomachean Ethics*. That is why he maintains that luck belongs to what is done by intention and choice, while chance belongs to these and to other things; these, although being done for the sake 25 of some goal, are to no purpose as far as that goal is concerned, but do achieve some other end which follows from the antecedents – whether this occurs in the case of natural or irrational or purposive entities. That luck belongs to the sphere of choice and rational activity he demonstrates from the fact that luck is found where good fortune[294] is found, and from the fact that good fortune is found in matters of rational activity and is to do with that activity itself. He further 30 demonstrates this point from the fact that good fortune amounts to the same thing, or nearly the same thing, as happiness, and that happiness is a sort of rational activity, and is successful activity. If, then, luck is found where good fortune is, and if good fortune is found where happiness is because it is in some way the same as happiness; and if happiness is found in rational activity because it is a sort of 346,1 successful activity, then luck is found in rational activity. But those who imagine that good luck *is* happiness are wrong. It certainly is one of the things related to happiness, but it is not the same or even similar, since good luck is uncertain, while happiness is something steadfast. This distinction is made in the *Nicomachean Ethics*.[295] 5

That luck belongs to choice and rational activity he shows also from the converse argument. For if where there is no rational activity there is no place for the outcome of luck, then it is clear that where there is the outcome of luck there is also rational activity, and that luck is to do with rational activity and choice. He reminds us that the outcome of luck has no place where there is no rational activity by 10 mentioning inanimate things, beasts and children. For these are said neither to engage in rational activity (because they do not exercise choice) nor to achieve anything as the result of luck. They enjoy neither good nor bad fortune, except by a sort of similarity or analogy. That is why even if Protarchus[296] says that 'the stones that form an altar are lucky because they are revered, while their fellows are trodden underfoot' we must understand what he says in the sense of 15 there being a similarity to good fortune. The similarity lies in the fact that the stones did not themselves join in contributing anything to their becoming altars, just as that fact[297] did not join in contributing anything deliberately towards the end achieved, an end which *we say* came about by luck for the stones without any intention on their part. Why, then, does the outcome of luck have nothing to do with them?

20 The answer is that such things cannot be the efficient cause of the outcome of luck, since luck arises as a consequence of (or subsists alongside) choice and rational activity, whereas they can undergo[298] something as the result of luck when the agent, to whom the outcome of luck does belong, achieves something as the result of luck in their case, for example the man who struck the tyrant Prometheus[299] in order to remove him, but who cut off his tumour and so cured him. The sword might in this way be called lucky and might perhaps even 25 be worshipped by the one who was cured; similarly an artist once painting a horse threw away his sponge, and froth issued by chance from the mouth of the horse in the painting – although that was not his intention.[300] And if somebody, chiselling a stone for some other purpose, by luck and chance created an altar, then the altar would have been created by luck. But perhaps these are not the lucky accidents of inanimate things, but of people engaged in activities 30 concerning them; for the painter and the sculptor were doing one thing, but something else happened. The term 'by similarity' applies equally to them, since what happened to them was different from what they were there for.[301]

197b13 But chance applies both to other animals and to many inanimate things. For example, we say that a horse comes by chance because by coming it reaches safety, not that it comes in order to reach safety. The stool falls by chance; for it is placed for someone to sit on, but it does not fall for the sake of this. So it is clear that, in the case of things that, generally speaking, happen for the sake of something, when something happens as a result of an external cause and not for the sake of what occurs, then we say that this is the result of chance. When something occurs as the result of chance in matters which are susceptible to choice and in the case of people who exercise choice, then we say that it is the outcome of luck.

35 Alexander writes '... both to irrational animals and to inanimate 347,1 things' – but the text does not denote anything different. This is a demonstration that chance has broader applications than the outcome of luck, since it is to do not only with humans, but also with other animals and also with inanimate things. This is confirmed by examples. For the horse which is captured by the enemy, and which, 5 growing thirsty, proceeds to its former haunts in order to drink and is recaptured by its owner, comes, we claim, by chance, i.e. it reaches safety by chance in coming, because although it reaches safety by coming, it does not come in order to reach safety. Similarly the stool which falls and which lands in such a way as to provide a seat is said to fall in this way by chance, since it falls so that it provides a seat,

but does not fall for the sake of this, namely in order to provide a seat, 10
but to occupy its proper place. Generally speaking, when, in the case of things happening for some end, some end is achieved 'as a result of an external cause and not for the sake of what occurs', and something other than the end in view is achieved, we then say that both these events occur equally as a result of chance. He is quite right to say that the end that occurs in this way must be externally caused 15
and not within the nature of the item, since it is the outcome of chance. For if the stone which falls were to be cube-shaped, then it would not be said to fall conveniently so as to provide a seat as the result of chance, since it is entirely within its nature to fall, if it does fall, in this way. So fire does not move upwards as a result of chance, since this is the inner moving nature of fire. Therefore he will say 20
that not even things that happen contrary to nature happen by chance,[302] since the cause – whether it be efficient or final or perhaps both – is not outside them, as is the case with what happens as the result of chance, when both the efficient cause and the end must be external and unpredictable. When a stone hits something, both the blow (which is the end, for it does not belong to the nature of the stone) and the efficient cause of the blow are external. Therefore the blow is a result of chance. Both these events are said to occur equally as 25
the result of chance.

He says: 'When something occurs as the result of chance in matters which are susceptible to choice in the case of people who exercise choice, then it is the outcome of luck.' He was right to add 'in the case of people who exercise choice', for it is by this that he distinguishes the outcome of luck from that of chance. For chance applies to 'matters which are susceptible to choice', even if the things in whose case it 30
occurs do not themselves exercise choice. For the horse's arrival is a matter of choice for the owner,[303] although the horse itself does not choose to come. The stool would be placed as it was as a matter of choice on the part of somebody, but it does not fall in this way by choice. Aristotle himself says, of the stone that strikes something by chance, that 'it might be caused by somebody to fall in order to strike 348,1
something' when that person having chosen that end throws it. It is for that reason, then, that such an action is in the field of choice and purposive. So events in the realm of purpose and the outcome of chance are simply what would have been chosen,[304] while events which depend on a choice are the outcome of luck. It depends on choice 5
not because what happens as a consequence was chosen (for then it would not be the result of luck), but because it supervened on a choice directed towards another end. Alexander says 'it is not always the case that, where something other than the object of choice is achieved, this is straightforwardly luck and the outcome of luck, but only if it is the sort of thing that would neither of necessity nor usually be

consequent on what occurs for the sake of something else. For when 10 a man takes too much exercise to build up his strength but grows feverish because of fatigue, then his fever is not an object of choice, but neither is it the outcome of luck; for fatigue usually occurs in most cases of immoderate exercise, and fever follows on from this. Similarly, when someone makes a voyage during a storm a shipwreck often occurs, not as a matter of choice, and certainly not as a result of luck.[305]

15 **197b22** An indication is the phrase 'to no purpose', which is used when that which is for the sake of something else does not occur so as to achieve it. For example, we take a walk in order to cure constipation; but if walking fails to achieve this end, then we say that we walk to no purpose, and that the walk is to no purpose, as if what is naturally done for the sake of something else is to no purpose when its natural goal is not achieved. For it would be ridiculous for someone to say that he had bathed in vain because the sun did not go into eclipse, since a bath is not taken in order to produce an eclipse. So chance lives up to its name when it happens to no purpose. For the stone did not fall in order to strike something, and so it fell as a result of chance because it might have fallen by somebody's agency and for the purpose of striking someone.

He has stated that we commonly[306] say that something happens as a result of chance when it is not the end in view that results from whatever happened with an end in view, but something other than 20 the thing, whose origin was not internal. To indicate that this is so he notes the fact that what is 'to no purpose' is such[307] because it runs parallel to what is by chance. In fact we say that something happens to no purpose when what is happening for the sake of something else does not attain the end for the sake of which it is happening. For example, if someone goes for a walk to cure constipation and fails to achieve that end by walking, we say that he is walking to no purpose 25 and that his walk is to no purpose; for what is naturally done for the sake of something else is done to no purpose when its natural goal is not achieved. Therefore if someone were to say that he has bathed to no purpose because the sun is not in eclipse, then he would be talking nonsense. For he did not take his bath in order that the sun should go into eclipse. The phrase 'to no purpose' has a two- or even three-fold application. Either nothing else follows what is occurring to no purpose, but the proper goal is merely missed, just as when 30 walking merely fails to cure constipation; or some result follows other than the one for the sake of which the action is performed; or else both the intended result and something else besides it follow. The stone fell not in order to strike something, but in order to find its proper place; so it fell in such a way as to strike something as a result of chance, while it fell *per se* in such a way as to find its proper place; 349,1 it fell to no purpose in relation to its striking something, and its blow was a result of chance; it might have fallen if thrown by someone in order to strike the person and so belong to the category of purpose, but it did not in fact fall for this reason on this occasion.

There is another reading of the beginning of this passage: 'An 5 indication is the phrase "to no purpose", which is used when the end is not achieved, but the means to it is.' This demonstrates the same point. For something is said to happen to no purpose when that for the sake of which some other prior event would have occurred does not happen as a result of that prior event, but when what happens is a consequence of something else – its *per accidens* cause. For when the blow does not occur because of some agent throwing <the stone> 10 for that purpose, it is said to occur as a result of chance, because it followed an event which preceded and which was to no purpose in relation to the blow. That was its movement to its proper place, which did not happen for the sake of the blow.[308]

I think it worth taking note of what Alexander says at this point: 'Aristotle does not mean that "chance" and "to no purpose" are the 15 same, as some thought; rather "to no purpose" and "chance" differ in the same way that "luck" and "the outcome of luck" do. For luck was the *per accidens* efficient cause, while the outcome of luck, which was subsequent to such a cause, was also a final cause of such an efficient cause. In this way, then, "what was to no purpose" would on occasions become the efficient cause to a chance event *per accidens*, and the chance event would be what is subsequent to what happens to no purpose. But it differs from the outcome of luck in that chance does 20 not always and in every respect follow on what is to no purpose, while the outcome of luck always follows on luck. But when it does follow, then the outcome of luck and the chance event have the same structural relation. Some people were deceived, as if Aristotle meant that "to no purpose" and "chance" are the same, by the unsuitable nature of the example which he gives. For he says "For the stone did not fall in order to strike something, and so it fell as a result of chance 25 because it might have fallen by somebody's agency and for the purpose of striking someone". For the example is inappropriate, since the stone might be said to have fallen to no purpose if the object of its falling did not come about[309] – this is what he said was to no purpose. Someone's being struck by it would have happened by chance if it were travelling for a different purpose but, failing to achieve it, struck 30 someone. We must supply to the words "so it fell as a result of chance" the words "and struck someone". For it struck him as a result of chance, but it fell to no purpose.[310] He showed that this is what he meant by adding "because it might have fallen by somebody's agency and for the purpose of striking someone". For he means that the 35 striking, not the falling, happened by chance.'

This is what Alexander says, word for word; I consider that it is well said that 'to no purpose' and 'chance' are not the same. For we say that that which has no consequence is to no purpose; we say that 350,1 the man who takes a walk to cure constipation has taken his walk to no purpose if he fails to cure constipation. But chance is that which something always follows, but only *per accidens*; so that the prior event itself occurs to no purpose only so far as the result is concerned, 5 and for that reason it is called chance. But that 'to no purpose' and 'chance' differ in the same way that luck and the outcome of luck do, and that 'to no purpose' is the efficient cause of chance, and that chance is the end and what follows does not seem right to me. For it is not that which is to no purpose which corresponds to luck, but chance, while the outcome of chance corresponds to the outcome of 10 luck. And Aristotle says that it is luck and chance, not what is to no purpose, that are efficient causes, and that the outcome of luck and the outcome of chance, not chance itself, that are final causes. At the beginning of the argument he says 'luck and chance are among the causes', and a little later 'therefore, as has been stated, both these – luck and chance ... are causes', and about ends 'every outcome of luck 15 is the result of chance' and again 'when any of these occur as the result of chance ... they are the outcome of luck'. But he neither says that what is to no purpose is the same as chance (but that 'chance' is derived from 'to no purpose', because the prior event occurs to no purpose only as far as the result is concerned), nor that it is the efficient cause of chance. But in as much as it is inefficient as far as 20 itself is concerned, <the intended result is not effected> – that is how far short it falls <of being an efficient cause>, which convinces us of his explanation that chance and what is to no purpose are such. In general what is to no purpose, in so far as it is to no purpose, is not even an efficient cause in the way that luck and chance are. For the end does not follow what is to no purpose, so that it is not even efficient *per accidens*; but chance, which is followed by the outcome of chance, 25 is efficient, since the prior event occurs to no purpose only as far as what follows is concerned, while the result occurs *per accidens*. But not even the example of the stone was inappropriately put. For just as he said that the horse arrives as a result of chance, i.e. is restored safely to its owner, so he says that the blow which causes the damage lands as a result of chance. Therefore the *per accidens* cause of the blow is chance, and the blow is the result of chance. The term 'to no 30 purpose' is applied not because the stone failed to attain its proper end, but because the fall occurred to no purpose in respect of the blow struck. For it fell not in order to strike a blow, but to find its proper place. If it had been thrown by someone in order to strike a blow and had landed, then the blow would not have been the outcome of chance. 351,1

Alexander shrewdly observed at this point 'and wanting to demonstrate that the outcome of chance is to be found in matters where choice is concerned and where final cause operates, he added "because it might have fallen by somebody's agency and for the purpose of striking someone".' For when one event (which follows after a prior event which is to no purpose as regards what follows) is of such a sort that something might have happened for its sake, although the prior event did not occur on that occasion for its sake,[311] then this has happened on that occasion by chance. Alexander says: 'it is clear that he calls what happens "chance"[312] from the fact that he ranks even the outcome of luck below chance and says that chance has broader connotations than the outcome of luck. For the outcome of luck is said to be the end of the process, and the survival of the horse is like this.' And he would be ranking the outcome of luck below chance when he says 'they differ in that chance has broader connotations'. But he ought to have taken what follows into consideration. For Aristotle adds 'every outcome of luck is the result of chance', just as even earlier he said that what is broader in scope than the outcome of luck, which he called chance, is the outcome of chance. For this corresponds to the outcome of luck. Speaking strictly he says that the horse arrived by chance, since the arrival is chance, while having arrived and found safety is the outcome of chance. If he sometimes uses the term 'chance' instead of 'the outcome of chance', it is hardly surprising. For Homer does the same when he says[313] 5 10 15 20

Menelaus of the loud battle-cry came by chance

and[314]

The gates of Heaven, guarded by the Hours, rolled open by chance

meaning that the result was chance, with no obvious preceding cause, and could more strictly be said to be the result of chance, on the analogy of the outcome of luck. For if the result is chance, what then would the outcome of chance be? Aristotle called the outcome of luck 'luck' when he said 'even so, in some instances we could ask ourselves the question: Could things that happen by luck be the causes of luck?' Plato too calls the outcome of luck 'luck' in the *Phaedo*,[315] when he says 'Echecrates, some luck has occurred', instead of saying that something has happened as the result of luck. 25 30

197b32 <The outcome of chance> is most distinct from the outcome of luck in the case of what comes about by nature. For when something occurs contrary to nature, then we say that it has occurred not so much as the result of luck, but rather of chance. But even this is somewhat different, since the cause of the one is external, of the other internal.[316]

It has now been stated what chance is, what luck is, and how they differ from each other.

He has stated that a distinction within the outcomes of luck and of
352,1 chance lies in the fact that some things are outcomes of both luck and chance, while the outcomes of chance are not all outcomes of luck. This distinction, embracing as it were whole and part through the association <of the less extensive> with the more extensive, showed that the outcome of chance included the outcome of luck. What is more, <we must take into account> those outcomes of chance which
5 seem to occur not as a result of luck, as is the case with irrational animals and inanimate things, since if they were events that involved choice they would have happened as the outcome of luck as well. For the horse would have reached safety as the result of luck too, if it got lost as a result of luck when it escaped from the enemy who had taken it to drink, and reached safety as a result of luck when its owner came out of the camp on some other errand; similarly, the stool fell by luck so as to provide a seat when it was thrown away by the thief who was being chased.[317]

10 Therefore he now tries to offer some other distinguishing feature which marks off the outcome of chance from the outcome of luck in the phrases 'what comes about by nature' and 'when something occurs contrary to nature'. For occurrences which are contrary to nature are not the outcome of luck since they do not belong to the category of things to do with choice; 'we say that it has occurred not so much as the result of luck, but rather of chance', he says. The reason why he
15 writes 'rather' instead of making no qualification he himself gives in the words 'but even this is somewhat different'. For what occurs contrary to nature will be said to occur by chance rather than as the result of luck. But in fact not even this is the result of chance in the strict sense. He again gives his reason for this, that in the case of what occurs as the result of chance the cause is external, and what occurs as a result of chance is a consequence of this as a *per accidens*
20 cause. For that which becomes the efficient cause (whatever it may be – we call it chance, and in the case of events where choice is involved we call it luck) of the stool's providing a seat by chance must be external to the nature of the stool. Of course with regard to what I said earlier, that if what landed were a cube,[318] then it would not be outside its nature to provide a seat, that would be the reason why it

could not then be said to occur even as the result of chance. For a consequence of chance cannot of necessity follow on a cause which is inherent and evident, since it would not be the result either of chance or of luck if the efficient cause, which produced a *per se* consequence, were evident. The term 'external', then, would indicate what is outside the nature of the thing involved, and it would also indicate something unclear and uncertain; and it would be more strictly accurate if it indicated both, since in the case of the outcomes of chance and of luck the cause should be neither inherent nor entirely evident. Yet in the case of what occurs contrary to nature the cause of its being so is something internal,[319] just as is the case with what occurs according to nature. For there is something which is *per se* the cause of what is contrary to nature. For pressure or imbalance or excess of matter or shortage or some other such thing is the *per se* cause of what is contrary to nature. Now he says that in the case of what occurs by chance the final cause is external; as he said earlier, this is not the efficient cause. For the blow <caused by the stone> and the safe arrival of the horse are events that supervene from outside. But in the case of what occurs contrary to nature the end is inside, not outside, itself; therefore this would not be the outcome of chance. In the case of what occurs by nature but not according to choice, whatever has its efficient and its final cause outside itself would not be the outcome of luck, but would be the outcome of chance, since everything which occurs not as the consequence of predetermined causes is in the broad sense the result of chance; but of these occurrences those which supervene on events subject to choice are the outcomes of luck, while those which are closely linked to a thing's nature are more specifically the outcomes of chance; by the term 'nature' we must understand all irrational life. He concludes his remarks at this point with the words 'it has now been stated what chance is, what luck is, and how they differ from each other'.

25 30 35 353,1 5 10

> **198a2** Each of them belongs to the type of cause involving the origin of change. For each is always either one of the things that cause by nature or one of the things that cause by intention. But there is an indeterminate number of these.

He has shown what luck is, what chance is, and how they differ from each other. He now systematically proceeds to tell into which category of the causes which he has enumerated he places them, saying that he places both of them under the efficient type of cause – for this is the type of cause 'involving the origin of change' – but in such a way as to claim that these are *per accidens* efficient causes. For this is why he adds 'but there is an indeterminate number of these' instead of 'but these are *per accidens*'. For causes of which there is an indeter-

15 20

minate number must be *per accidens*, since the number of *per se* and not *per accidens* causes is determinate. He bases his proof that these are efficient causes on the fact that the *per se* causes, on which these
25 are contingent, are efficient causes. 'For each is always either one of the things that cause by nature or one of the things that cause by intention', together with which luck and chance subsist. Nature and intention are efficient causes. For if the stone when dropping hits someone because of its weight, its nature is the *per se* efficient cause of its dropping, while chance is that of its striking someone. If someone digging a trench for foundations finds a treasure, the inten-
30 tion is the *per se* cause, while luck is the *per accidens* cause, when he is digging not in order to find treasure but to make a trench or something else. So if luck and chance are *per accidens* causes in the case of things that happen for the sake of something, and if of these things – the things that occur by nature or those that occur by intention – there is always some *per se* cause, and if these are efficient causes, then it is clear that those things too, which subsist together
35 with them, would be efficient causes.

354,1 **198a5** But since chance and luck are causes of things whose causes might be mind and nature, when something becomes a *per accidens* cause of these very things, since nothing which is *per accidens* is prior to what is *per se*, then it is clear that no *per accidens* cause is prior to a *per se* cause. Therefore chance and luck are posterior to mind and nature. So however much chance might be the cause of the heavens, mind and nature must of necessity be prior causes of many other things and particularly the universe.

He has shown that luck and chance, even if they are causes, are *per accidens* causes, and that they are *per accidens* causes of the things
5 whose *per se* causes are mind and nature, since this has been demonstrated in the case of what happens for the sake of something. He assumes that 'nothing which is *per accidens* is prior to what is *per se*', and for this reason that 'no *per accidens* cause is prior to a *per se* cause' (for a *per accidens* cause is one of the *per accidens* entities). But what is *per accidens* is in no way comparable to what is *per se*, since the
10 former supervenes on the latter. From this he concludes that mind and nature are prior to luck and chance. The conclusion of the argument is this: mind and nature are *per se* causes; *per se* causes are the primary causes of *per accidens* causes, and the primary causes of *per accidens* causes are the primary causes of luck and chance; therefore mind and nature are the primary causes of luck and chance.
15 But since *per accidens* causes are posterior to *per se* causes, and since the latter pre-exist and the former supervene on them, it is clear that

where there are *per accidens* causes, *per se* causes must pre-exist. Therefore where luck and chance are *per accidens* causes, there mind and nature must pre-exist. Consequently, for those who think that luck and chance are the causes of the heavens and the whole universe 20 it follows from what they say that they must posit mind and nature as the primary causes of these. 'That is why,' as Alexander says, 'the primary causes of the primary entities in the universe are efficient causes, and the primary entities in the universe came into being because of the primary causes, namely mind and nature. If in general mind and nature are the causes of some things, then they would be 25 the causes of such things as are inherently rational, such as the heavens and the universe.' But perhaps we do not need the primary causes of the primary entities in the universe to be efficient causes in order to strengthen our belief that the heavens came into being as a result of mind and nature. For this advocacy needs another which shows that the heavens are the first of all entities. Yet the argument of Aristotle's is self-advocating, which shows not simply that mind 30 and nature are causes prior to luck and chance, but that where luck and chance are causes, there mind and nature are the causes prior to them, *per se* with what is brought to completion *per se, and per accidens* with what is brought to completion *per accidens*. For going out to meet a friend is the cause of greeting one's friend and running 355,1 across one's debtor; it is the same cause, but the *per se* cause in the former case, the *per accidens* cause in the latter, when it is said to be luck, and, in the case of natural things, chance. Again, mind or choice might be the cause of the same result both *per se* and *per accidens*. For one might find one's debtor in the market place from choice, and 5 the horse might reach safety because of someone's foresight.[320] So the heavens, if they are the product of luck and chance, would be the product of mind and nature in a considerably prior sense.[321] He says 'of the many other things and particularly the universe', because he wants mind to be the cause specially of things that are eternal, since they are ordered and defined. This universe is eternal, as he showed 10 in *On the Heavens*, as are many of its contents – heavens, stars and the sum totals of each of the elements.[322]

On this passage Alexander comments: 'What Plato wants to show and does extensively show in the tenth book of the *Laws*,[323] Aristotle 15 here demonstrates briefly, by defining just what the nature of luck is, and showing that it is posterior. For what might have occurred even as a result of mind and nature is said to have occurred as a result of luck and chance when they become one of the *per accidens* causes.' We should realise that Plato does not show that luck is posterior to mind and nature, but that luck and nature are posterior to soul, filling out the doctrines of the ancient natural scientists, those who talked 20 in particular about the material causes as being the ones that impinge

first on our consciousness, having as yet nothing to say about efficient causes. Plato remarks, as I said, not that luck is posterior to nature and mind, but that soul and art are prior to nature, writing as follows:[324] 'Let me put it more plainly still. Fire and water, earth and
25 air – so they say – all owe their being to nature and luck, none of them to art; they, in turn, are the agents, and the absolutely soulless agents, in the production of the bodies of the next rank, the earth, sun, moon and stars. They drifted casually, each in virtue of their several tendencies; as they came together in certain fitting and
30 convenient dispositions – hot with cold, dry with moist, soft with hard, and so in all the inevitable casual combinations which arise from blending of contraries – thus, and on this wise, they gave birth to the whole heavens and all their contents, and, in due course, to all animals and plants, when once all the seasons of the year had been produced from those same causes; not, so they say, by the agency of
356,1 mind, or any god, or art, but, as I tell you, by nature and luck. Art, the subsequent late-born product of these causes, herself as perishable as her creators, has since given birth to certain toys with little real substance in them.'[325] In addition he shows[326] that 'our soul was
5 created prior to our body, and body is secondary and posterior to soul, the latter ruling the former according to nature', and concludes 'we remember agreeing earlier on that if soul were to be obviously older than body, then the attributes of the soul would be older than those of body'. It is clear, then, from his words that Plato's argument in his interpretation of the ancient natural scientists here ranked luck
10 alongside nature. He shows that soul is prior to both of these, requiring a lengthy discussion on this point in order to prove that the self-moving essence, viz. the soul, is the source of all movement. Plato seems in this passage to be calling luck the fashioner of the combination of the natural powers with each other, but without any evident cause. This in my opinion is the meaning of the words 'they drifted casually,[327] each in virtue of their several tendencies; as they came
15 together in certain fitting and convenient dispositions ...'. Eudemus shows that nature is prior to design, and that design is prior to luck.[328]

When considering this passage of Aristotle, it is also worth taking into account the views of those who think that Aristotle does not mean the efficient cause of the universe. For if 'mind and nature must of
20 necessity be prior causes of the universe and of anything else' and a cause in the way that chance and luck are said to be by some, and if these 'belong to the type of cause involving the origin of change', i.e. if they are ranged among the efficient causes, it is clear that mind and nature would be efficient causes of the universe. I know that
25 someone of a contentious turn of mind might say that the argument that if luck and chance are the cause of the heavens, mind and nature would be prior to them, is based on an assumption. But we must

remember Alexander's words, that the primary causes of primary entities would be efficient causes. Mind and nature are primary causes, and the primary existents – or rather things that come-to-be in any way whatsoever even if not from the beginning of time[329] – are the heavens and the universe. 30

So far I have paid as close attention as possible to what Aristotle said about luck and chance, taking each point in turn; it would therefore be sensible to summarise briefly Aristotle's views on the whole subject, and in this way to add the opinions of more recent thinkers and to show that they differ in no respect from ancient tradition. 357,1

Since Aristotle, in his search for the causes of what occurs in the natural world, finds that some things are said to occur as a result of luck and chance, he proposes to articulate what these terms signify according to current opinion. Since, then, there must be an efficient cause of everything that comes-to-be, and since every final result must be preceded by something on which that which comes-to-be supervenes; and since some of these act in advance for the sake of what comes-to-be, and since that which properly comes-to-be is consequent upon what acts in advance (as when someone goes out to meet his friend and actually does meet him, or when a stone falls in order to occupy its proper place, and does occupy it); but since sometimes what supervenes does not follow as such on what has been done beforehand, and sometimes what has been done beforehand does not end up with the result for which it was done, but with some other one which turns out otherwise and not according to the original end in view of the activity – as when one goes out to greet one's friend, and this doesn't happen but one gets one's debt back instead – then there is some other end supervening and not the one for the sake of which the action was performed. When the stone which drops in order to occupy its proper place becomes a seat, it attains some end consequent upon its fall other than the end aimed at by its fall. All such events where some end supervenes upon the previous activity other than the one aimed at are said to occur as a result of chance; they occur as a result of the prior activity, since they would not have occurred had it not been performed (had you not gone out, you would never have found your debtor in the market place, and had the stone not fallen the seat would never have been formed); but when the initial activity does not achieve the consequence for the sake of which it happens and for this reason is itself to no purpose, and when it does not occur for the sake of what actually does follow and for this reason again it is to no purpose as far as the consequence is concerned, then it is quite rightly called chance on the grounds that it itself is performed beforehand to no purpose[330] in regard to such a result. Even so, it is still called a cause, because the consequence would never 5 10 15 20 25

have occurred if it had not preceded, and because we can see no other cause if we think like natural scientists. But this is not the *per se*
30 cause of such a result, but *per accidens*. They say that those of the outcomes of chance that supervene on actions done by choice happen as a result of luck, and that luck is the cause that subsists alongside choice and activity concerned with choice, such as going out to meet one's friend, when not the result that was intended follows, but something other than it; on the other hand as far as the remaining outcomes of chance are concerned, which supervene on irrational and
35 natural activities, using the name of the common characteristic, they call these too the outcome of chance, and they call the antecedent
358,1 cause chance, which is followed by the result not *per se*, but *per accidens*. Such, then, are the views which Aristotle propounded aimed at articulating his usual terminology according to the measures proper to his scientific teaching about causes.

5 I think it is worth first noting that perhaps the ancients claimed that not only what supervenes on changes involving choice was the outcome of luck, but also what was consequent upon natural deviations. So that I may pass over the majority of their treatments of this topic, some of which I have set out above in my discussion of Empedocles, the treatment given by Aristotle is sufficient. Speaking about
10 the air, Empedocles says: 'For sometimes it chanced to be running in this way, often in another.'[331] Plato too, researching into the opinions of earlier natural scientists about the outcome of luck in the elements and the mixtures of the elements says that they put the responsibility on luck when he writes: 'Fire and water, earth and air – so they say
15 – all owe their being to nature and chance', as well as in the passage just after this which I quoted a little earlier.[332] On many occasions the ancients call what supervenes on choices chance and the outcome of chance, for example the line of Homer: 'Menelaus of the loud battle cry came of his own accord.'[333] For Menelaus' arrival supervened on the assembling of the other chiefs by Agamemnon, and that occurred
20 by choice. If, then, they place luck and chance under the same heading without distinguishing them, not taking luck in the sense of 'happening upon', but in the sense of 'just as it happened'[334] (when they are unable to find a cause), and taking 'chance' obviously in the sense of 'causeless' (for when there is no defined cause, then we usually talk about 'chance' and 'just as it happened'), then the phrase 'it chanced
25 to be' bears the same sense. This sense too is taken from 'happening upon', but happening upon not according to reason or a preceding goal, but randomly and anyhow; for that is what 'just as it happened' signifies. This is the distinction I draw between the two terms.

30 Further, I think it worth asking whether luck is only applicable to matters where choice is involved, and whether it is only applicable to such instances when they happen less often. For since 'happening

upon' and 'hitting upon'[335] are said to be exclusively the outcomes of luck, and since we say that even the best marksman 'hits' the target, and the best artist 'hits upon' <a good likeness> and the poor artist fails to do so, then luck is applicable to what happens more, not less, often. Eudemus says: 'If one accomplishes something by one's skill, 35 then it is called hitting the mark, but otherwise it is called missing 359,1 the mark.'[336] When someone proposes to do something, we say that he hits the appropriate mark when he achieves his aim, just as we say he misses it when he fails to achieve it. Therefore hitting the mark is not *per accidens* when the man proposing the action succeeds. If, then, we think that certain divine causes pre-exist as in the case of all other specific qualities such as beauty, health or victory, in which 5 participation is granted to the participants, and if we are prepared to call these causes by the names of the benefits granted by them, since attaining[337] the benefit which presents itself is something remarkable and worthy of the divine gift, then why are we not bound to call the divine goodness responsible for the attainment 'luck'? So some approve of Aristotle's words: 'luck is a cause, but it is unclear to human 10 thought, since it is something divine and rather numinous.'[338] But if we say that luck is a cause particularly in matters in which we can see no other *per se* cause recognised, we ought not for that reason to think that it is the *per se* cause of anything when it is the *per accidens* cause of something else, and then call the cause luck and the result the outcome of luck, but think that the *per se* cause is the cause of what comes-to-be. For example, going into the market place to meet 15 one's friend is the cause of one's being in the market place, and the choice and the arrival are joint causes[339] of meeting one's debtor, but the cause in the strict sense is luck, which meant that he would have met his debtor even though he arrived for the other reason; but the choice too was on that occasion a cooperating[340] factor; and if he had not come for that reason, luck seems to be the only cause; but on that 20 occasion choice was altogether a cooperating factor in one's arrival, but since one did not arrive for that purpose, it clearly lacked the guiding cause. Yet Aristotle, as I have said several times, making a natural scientist's distinction, left it to the theologians to posit the uncertain cause, while he himself calls the recognised cause 'luck' (when it achieves some end other than the one proposed), and he calls 25 the result the outcome of luck.

When we ask where the realm of luck lies, we shall discover that it is where there is a need for a lucky happening. Things that need to participate need something; things distinct from each other participate. Consequently, even in the distinctions between intelligible 30 forms there is need of luck, in order that what are distinct may achieve participation in each other. But if that distinction lacks distinction, and if the participation is not participation but rather consubstantiality, it is clear that the particular nature of luck is never apparent among them as it is in the material world, where complete distinction and separation, however it happens, is already a fact; in it both participation and hitting the mark are evident, and luck displays its power more clearly. For the sun and each of the planets reaches†[341] its position in each of the signs of the Zodiac through luck, as they do their configurations in relation to each other, as the moon gains† the light of the sun, and as all the stars in the heavens receive† each other's rays. But the power of luck is not so evident in the heavens because of the necessary nature of the organisation; but in the sublunary sphere, where there is a strong possibility of not achieving† because of the concatenation of many indeterminate causes, there in particular chance displays its mastery; it channels all the causes so that each thing should not miss but achieve what is rightfully presented to it, that is to say what each thing deserves. The goddess Health is particularly evident where diseases strike, and even more evident where more particular causes are absent.[342] We say that justice is the cause of fair distribution, and that luck is the cause of getting† what one deserves. It is particularly in evidence when neither intention nor any other obvious cause is apparent. There are cases where luck is responsible for causing other causes to achieve† their end,[343] and it is especially evident where there is no other recognised cause; and when there is no particular or apparent cause for something, it happens as a result of luck (as is the case with this goddess[344]) and from chance. Therefore luck does not operate exclusively in what happens less often (as, for example, the children of the wealthy are usually themselves wealthy and blessed with good luck), nor is it a *per accidens* cause. For it is the cause in the strictest sense of each person's gaining what is presented to him. It presses into its service all mortal things[345] as causes, both *per se* and *per accidens*. If they claim that luck operates in what is disorganised and less often the case, we will not accept some of what they say (for things in this world are not altogether disorganised, but share some measure of organisation; nor does chance operate only in what is less often the case, as has been said); but we will approve of some of their other claims. For the dominance of Luck (*tukhê*) controls the sublunary part of the universe, which is the realm of the nature of the possible, which, being disorderly in itself, Luck controls, organises and guides along with the other primary causes. That is why they represent her with a rudder to guide with, as if steering the vessels on the Sea of Becoming, and they fix the rudder onto a sphere to control the constant flux of Becoming. There is a Horn of Plenty, full of fruit, in one of her hands, the cause of achieving† all the divine fruits.[346] That is why we honour the *Tukhê* of every city, home and individual, because if we stand far away from union with the divine, we are likely to miss the participation which presents itself. In order to succeed† we need the goddess Luck and the deities which have the same characteristic among important nations.[347] All luck is good; for it is the full attainment† of some benefit. For no evil emanates from the divine. Of benefits some are attractive, others are punitive and corrective – and we usually call these evils. That is why we usually give a good name to the luck that is responsible for our achieving positive benefits, while we label as bad the luck that causes us to suffer punishment or correction. In the *Laws*[348] Plato presents this luck in its entirety as ordained by the Demiurge when he says that 'God controls everything, and, together with God, Luck and Occasion control all human affairs'. But let us proceed to the rest of Aristotle's treatise.[349] 361,1 5 10

<CHAPTER 7>

> **198a14** It is clear that there are causes, and that there are as many of them as we say. For the question 'Why?' embraces just this number. For the question 'Why?' can either be ultimately reduced to 'What is it?'; this is the case with unchanging entities, for example with mathematicals, where one can ultimately reduce the question to the definition of the straight or the commensurable or suchlike. Or it can be reduced to the first mover; for example, 'Why did they go to war?' 'Because the enemy invaded.' Or to the question 'For the sake of what?'; 'In order to dominate.' Or, in the case of things that come-to-be, to the matter.

He has conducted an investigation into causes – what they are and how many there are of them – which was obviously his original proposal. He has shown that they all fall under the four most obvious types, and that luck and chance can be brought under one of them, the beginning of change.[350] Since it is a considerable task to reduce so great a number of particular causes to so few types, he quite reasonably considers that his argument needs a word of explanation, and he shows in effect that the causes are just so many in type in the following way. Causes are the answer that we give when asked the question 'Why?' When asked this question we give four types of reply: the form, the agent, the end or the matter. These, then, are the four causes. He explains how each is reduced to such a type by examples. First, the question 'Why?' can ultimately be reduced to the form, as in the case of mathematicals. He calls these unchanging in contradistinction to natural entities, whose essence lies in change. 'Why is this straight?' 'Because it is the shortest of lines that have the same ends.' 'Why is the side of the hexagon commensurate with its diameter?' 15 20 25

'Because they are measured in the same units as the distance from the centre.' And when we are asked 'Why are the radii of a circle all 30 equal?' we will answer 'Because the circle is a plane figure embraced by a single line, and all of the straight lines inscribed within the figure from a single point[351] and reaching it, the circumference, are equal to each other.' Nobody who has stated this looks for any further cause.[352] Secondly, when we are asked 'Why did the Thebans go to war with the Phocians?' we will answer 'Because the Phocians plundered their temple', so giving the efficient cause. Thirdly, when asked 'Why did 35 the King of Persia attack the Greeks?' we will reply 'Because he 362,1 wanted to dominate the Greeks', giving the end and the goal. Fourthly, in the case of material entities which he calls 'engendered' (for plants, animals and their parts are engendered, and all these are material), we often reduce the question 'Why?' to the matter. 'Why did the foundation rot?' 'Because it was made of wood.' 'Why does the 5 man have <only> four fingers?' 'Because the matter ran out' (we say). 'Why is the engendered thing a human?' 'Because it is the product of sperm and menstrual fluid.' It is possible too to reduce the explanation to all these causes in the case of the same entity. 'Why does a man die?' 'Because he is a rational mortal animal', or 'Because what gives him his being comes and goes' or 'Because he is made up of 10 perishable matter' or 'Because it is better for him to die.'

Porphyry says 'It is possible to answer the question "Why?" in four ways; but the question "For what purpose?" can be asked only where there is a final cause: "For what purpose did they go to war?" "In order to dominate".' Alexander reconciles what he said earlier, that mathematicals do not have a final cause, with what Aristotle says in this 15 passage, that 'the question "Why?" can be ultimately reduced to "What is it?"; this is the case with unchanging entities, for example with mathematicals.' He himself explains it by saying 'Since in the case of mathematicals the goal is nothing other than the form (for in their case this is the final and ultimate cause), the form and the definition are what are ultimate in their case'. I have already stated 20 and I will now reiterate it, that the mathematicals do have an end.[353] They were revealed so that the soul could pass to the intelligible from the sensible; but such an end as this is not conceivable for mathematicals *qua* mathematicals, just as natural scientists do not have to proceed from natural science to theoretical philosophy;[354] so they quite understandably reduced the question 'Why?' to the last defini-25 tion; for they take the definition as the starting point, and they are unable to go beyond the starting point.[355] But since what comes-to-be has different features attributable to different of its causes – to matter, to form, to agent or to end – then when we are asked 'Why?' we quite reasonably refer to the appropriate cause. And we often give different causes even in the case of the same thing. 'Why am I sitting down?' 'Because my body contains a lot of the earthy and a lot of the fiery, which is why it lies down and gets up again':[356] that is the material explanation. 'Because my limbs are so jointed': that is the formal explanation. 'Because my soul chose it to be so': that is the efficient explanation. 'Because I thought it better so': that is the final explanation. But when someone is asked 'Why is Socrates' portrait snub-nosed?', to what else will he refer the cause than to the model?[357] I do not imagine he would relate it to the matter, or to the form, or to any other of the four causes, but would say 'Because Socrates was snub-nosed'. So that if we relate the question 'Why?' to the model too, the model too would be a cause. That is why Plato related the cause of the universe's being unique to the model, saying 'Were we correct in saying that the heavens are unique, or was it more correct to say that there are many heavens? They are unique, if they are to be crafted in the likeness of their model.'[358] In general, if enmattered forms are participations in primary forms, they are images fashioned in their likeness. Every image, then, is related to its model. Earlier I tried by means of a division to discover the necessity for the causes to be as many and of the kind that they are, and so it would be pointless going over the same ground. But the following point should be added: a natural entity which comes-to-be is entirely made up of form and matter; but only when it is coming-to-be does it come-to-be by some cause because of some benefit, and not when it actually is. Therefore everything that is of this kind is so through matter, form, agent or end. Since Alexander quotes '... or for what comes-to-be to the matter', we should pay attention to his explanation; for he says 'Aristotle said this either in contradistinction to what is ungenerated, since there is no matter in eternal entities, or because it is possible to find matter only in the case of things that properly speaking come-to-be, and these are substances. For in what comes-to-be and is *per accidens* there is, *per se*, no matter, but matter is in them by reference to substance.'

30

363,1

5

10

15

20

198a22 Since there are four causes, it is the natural scientist's job to know about them all, and as a natural scientist he will refer to them all when answering the question 'Why?' – material, formal, efficient and final. Often the last three can be subsumed under a single designation. For what something is and what it is for are often the same, and the primary origin of change is often the same in form as these. For a man begets a man, and this is generally the case with what is changed while it changes. Things that are not like this are beyond the remit of natural science, since they change other things without having change or the source of change within themselves, but are unchanging.

Thus there are three sciences – one concerning the unchanging, one concerning what is changed but is imperishable, and one concerning what is perishable.

He has said that there are four causes of natural entities, since causes 25 are the answers we give to the question 'Why?' He proceeds to state that the natural scientist 'will refer to them all when answering the question "Why?", speaking as a natural scientist ought'. For it is possible for a scientist not to answer in a scientific manner, just as it is possible for the geometrician not to answer in a manner befitting a geometrician, when they are arguing from principles not proper to their field. Since, then, the natural body embraces both matter and 30 form as well as an efficient cause since it is generated, and since the agent formulates the end, it is clear that the scientist must know all the causes of the natural body and relate the question 'Why?' to each of them appropriately.

Aristotle says 'Often the last three can be subsumed under a single designation' – the form, the end and the origin of change. This 364,1 observation, that three of the four come to the same thing, is pertinent to the person discussing the number of causes. Yet the scientist will find it useful to know this when relating things to the causes, so that when he tries to relate the question 'Why?' to the form or to the end he will not always hunt for one thing and then another to relate the form to, because often form and end are one and the same. For nature 5 contrives all antecedents for the sake of the form. These two things are one and the same with regard to number and the substrate, but two in definition. For if the ripening of the fruit consists in enabling the seeds in the pod to produce another similar plant, that is *tout court* both form and end. But as for the origin of change and the fact that something the same as these has been produced – they are not the same in number or according to the substrate, but only in form. 10 'For a man begets a man', and the father, being the same as the offspring in form, acts as a cause for the offspring. He says 'this is generally the case with what is changed while it changes', i.e. natural entities which have the source of change within them and are agents, since the discussion is about that sort of efficient cause. If, then, these are the same as what comes-to-be, they are the same in form.

15 But 'things that cause change without themselves being changed' are entirely outside the scope of natural science and belong to theology. For they do not have the source of change within themselves, since they are not changed. So unmoved causes do not belong to natural science. Nor are they a source of change in the way that nature is, for nature is changed while it causes change; so it is not within the scope of natural science to talk about these things. But 20 perhaps he is using the words 'this is generally the case with what is

changed while it changes' to show his opposition, albeit distantly, to those who think that the Forms are causes of the entities in this world being of the same form as them, but <unlike them> unchanging. For that which is of the same form bears the same name in the same sense, but these entities bear the same name but in a different sense, as deriving from a single source.[359] This, then, is how one might explain it, if he were to think that all efficient causes were the same as formal causes because of the form, if they are changed while causing change. But the sun while in motion causes motion and is active, but it is not 25 the same as the things it affects. For not even the things affected by it are all like each other in form, except, of course, that the form, in so far as it is enmattered, is both agent and product, while the matter is commonly said to belong to the substrate. For even if according to Aristotle there were no matter in the heavens on account of the fact that it does not change from some things into others, nevertheless there is a substrate in those things in that they change their places, 30 as he says in the first book of the *Metaphysics*.[360] Perhaps the person constructing arguments according to those in that book is finding similarities of arguments. But it is possible to understand the words 'this is generally the case with what is changed while it changes' in the sense of proximate efficient cause in the terms of natural science, namely nature. For nature has much in common with the form, as has been argued above. For the sun is not a proximate cause, even if 365,1 it changes things while itself in motion. Perhaps the conclusions fit in with this. For the causes which change and affect are threefold, some being unchanged, others being changed, of which some are proximate, others remote; he demonstrates this from the fact that the science of existing things – or rather the principles and causes of existing things – can be divided into three fields. One is concerned 5 with the cause which changes other things while remaining unchanged itself, i.e. metaphysics; another is concerned with causes that are changed but remain imperishable, i.e. astronomy; and the third is concerned with things that are entirely subject to coming-to-be and passing away and are themselves changed, namely the proximate causes – these are the studies of coming-to-be and passing away, of meteorology and animals and plants, and are the studies which 10 enquire into this sort of cause. It is clear that the proposed study which enquires into the principles of natural entities generally conducts its researches into causes which are themselves subject to change.

It is worth noting how he says 'the primary origin of change (i.e. the efficient cause) is the same in form' – but not in substrate and 15 number, in the way that the final and formal causes are related to each other. For nature was shown earlier as being an efficient and a formal cause, but perhaps being one also in number is inherent in nature along with being the form, while being the same in form[361] is inherent in an unqualified sense in all proximate efficient causes. As 20 a result of this I think it can be seen that Aristotle knew that the unchanged and primary cause was not only final but also efficient. For the primary origin of change, i.e. the efficient cause, he divides into the unchanging and the changed, saying that what changes while itself being changed is the subject matter of natural science, but that what changes other things while remaining itself unchanged is not the subject matter of natural science, taking 'causing change' in the 25 sense of affecting. On this passage Alexander says: 'It is the job of natural science to discuss the efficient cause, but not every efficient cause, but only those which are themselves changed when they change and affect other things, as if Aristotle were saying "in general those efficient causes which are themselves changed", since clearly according to him there are unchanged efficient causes'.

30 **198a31** Therefore the answer to the question 'Why?' is reduced to the matter, the essence and the primary cause of change. For people usually think of the causes of coming-to-be in this way, by asking what followed what, what affected first and what was affected, and in this way they always proceed to the next stage.

If the final and the formal cause often come to the same thing, numerically speaking, it is quite reasonable to reduce the answer to the question 'Why?' sometimes to the matter, sometimes to the form, which is the same as the end, and sometimes to the efficient cause. 35 For to leave out the final cause when listing is indicative of this.

366,1 At the same time it seems sensible to me to demonstrate from a more detached point of view the reason why the natural scientists[362] omitted the final cause. Because the form and the end seemed to amount to the same thing, they were content to talk about the form. That is why he immediately proceeds to point out why in their consideration of the causes of coming-to-be they reduced the causes 5 to these. For when they ask what happens after what, they are seeking the formal cause (for that is what comes-to-be), and when they ask what affected first they are in pursuit of what properly speaking is the efficient cause, just as when they are asking about what is affected they are dealing with the material cause. For matter is that which is affected by an agent, and the form is the engendered 10 affection. This can be indicative of the fact that the natural scientist in his researches reduces the causes to these questions in particular – what affects first, and what affects subsequently, what is affected (i.e. the matter), and what comes-to-be?[363] These concerns to do with natural entities are proper to the natural scientist, while first philosophy is to do with the unchanged cause, the immaterial and unaffected essence which is apart from coming-to-be; all these stand in contrast to what has been discussed above.

Alexander says: 'The phrase which reads "the causes of coming-to-be" can follow what has been said earlier, where he writes "Since there are four causes, it is the natural scientist's job to know about them all, and as a natural scientist he will refer to them all when answering the question 'Why?' – material, formal, efficient and final"; this could be followed by: "For people usually seek the causes of coming-to-be in this way", with the words between interpolated to define the identity of the causes and the difference in the efficient cause, since one sort of efficient cause is unchanged, and another changed, this latter being further divided into the one that is imperishable, and the other that is perishable.'[364] He can write 'and in this way they always proceed to the next stage' because the shoot proceeds from the seed, and then the stalk, and next the ear; but this can, as Alexander says, be deduced, because after the primary agent they move on to the proximate agent, and after the primary matter to the proximate matter in similar fashion, not proceeding to any other type of cause – the final cause – and the more so because they are looking for the proximate in the case of each cause. 15 20 25

> **198a35** But there are two principles of natural change, one of which is not natural, since it does not have a principle of change within itself. Anything which changes other things without itself being changed is like this, such as what is entirely unchangeable, the first of all things, the essence and the form; for that is its end and what it is for. 30

He already earlier on made a division within the efficient cause when he said that some things cause change while themselves being changed, while others remain unchanged. He now makes the same division with the words 'there are two sources of natural change' adding 'natural' in contradistinction to the products of art and choice. 35
Perhaps he picks up this division within the efficient cause in order 367,1
to distinguish completely the primary cause of change from the things that cause change by art or choice. First of all he draws a distinction between what is itself changed while causing change and what remains unchanged, and he makes that which causes change while itself remaining unchanged the primary cause of change.[365] But now, since art and choice, which are unchangeable (since they do not 5
increase, become altered or change their position), cause natural changes, he first distinguishes art and choice from nature and the primary cause by the fact that the latter pair cause change naturally (i.e. they cause change from within) while the former pair do so from outside. In the following way he distinguishes the prime mover of

change, on the one hand, from nature, or rather the things that cause
10 change according to nature (for nature does not *per se* cause change, but with the help of the body in which it resides); and he distinguishes it from these latter in that they have the principle of change within themselves, while it does not. For this is not a natural principle, but a principle in what is natural; if it had a principle of change within itself, it would be natural. On the other hand he distinguishes the primary principle from what causes change according to art and
15 choice by the fact that, even if *per se* they are not subject to natural change, they are changed *per accidens* within natural entities. For such are the activities of the soul which change their place and alter from one state to another in the process of exercising choice and art, while 'the first of all things' is 'entirely unchangeable' both *per se* and
20 *per accidens*. It differs from what causes change according to art and choice in so much as it is the first of all things. 'What is entirely unchangeable, the first of all things' is not like them; rather, the difference between it and what changes without being changed demonstrates this in both respects.[366] A little later he will show that it is first in every sense of the term. If, then, we are right in claiming that
25 this is the point of Aristotle's remarks, the distinction within the efficient cause will not have been made to no purpose. Since there is this division within the efficient cause, as Alexander agrees there is, then it clear that Aristotle thinks that the first of all things is also an efficient cause. But if there is a division only within the moving cause, then the first of all things can change as an end too.[367]

Having stated that the first of all things is itself unchanged, he
30 indicates that the form and the shape too are unchanged when taken as 'the end and what it is for'; for the end must be unchanged and determinate, since what is changed is changed for some end. If it itself were changed, not even the agent aiming at it would remain constant.[368] The form is not an end when it is taken just as it happens to be, but when it is taken in a particular way because it is better so.[369]
35 Alexander says: 'In this way he distinguishes mathematical form from form in natural entities, because the latter is aimed at also as
368,1 an end and a goal, because it is better so; for a form such as "straight and not otherwise" is better so, while in the case of mathematicals no such thing is sought. It would be stupid to try to demonstrate that it is better for the circle to have radii of equal length, and for the rectangle to have sides which encompass it.' Perhaps if one were to
5 understand the definition of the circle or the rectangle in the same sense as that of a man, not according to the shape but according to the potential, then he would discover that the shape was as appropriate to such an account as 'straight' is in the case of a man.[370]

> **198b4** So since nature is for some end, it too should be known. We should answer the question 'Why?' in all its aspects, e.g. that this is the necessary result of that (either simply or generally speaking); and whether this is likely to be the case, as the conclusion is of the premisses; and that this is the essence; and that it is better so – not simply, but in relation to each essence.

Since the natural form is an end and an aim, it is clear that the nature which creates the form creates everything for the sake of the form. If, then, nature creates for the sake of something and not to no purpose, the natural scientist *qua* natural scientist must know the nature of what comes-to-be, and he must be able to give a complete answer about the causes. So if we put the comma after 'is for some end', then the word *kai* should be taken as a conjunction with the phrase 'in all its aspects' <reading: 'So since nature is for some end, then this *too* should be known, and we should ...'>. If this is not the case, as we read in some versions, we should put the comma after 'this should be known' <reading: 'So since nature is for some end *and* we should know this, then ...'>, so that the reason for giving an answer to the question 'Why?' is to be understood as being the need to know that nature is active for the sake of something. The causes we must give are the efficient and the material. He indicates this by writing: 'that this is the necessary result of that (either simply or generally speaking)'; for among the eternal and divine entities the agent acts simply, since such-and-such a configuration of the moon always produces such-and-such moonlight. But in the world of coming-to-be it is generally the case, since sleep generally speaking causes growth when nature is dormant. By writing 'and whether this is likely to be the case, as the conclusion is from the premisses', he reminds us of the material cause, which is a *sine qua non* of an argument; for if a certain conclusion is to follow, then certain premisses must be laid down, which are described as the matter of the syllogism. So if anyone were to ask why such-and-such a conclusion had been drawn, we would refer him to the matter in answering the question 'Why?' That is the reason why the premisses are such. Similarly, in questions of natural science, if a particular form is to be brought into being by nature, matter of a particular kind must be presented beforehand; if, for example, a boat is to be fashioned out of a certain type of timber, the timber must be procured first if the boat is to sail well and be properly put together. By writing 'and that this is the essence', he is pointing out the formal cause, and by the words 'that it is better so' he is reminding us of the final cause. For both Plato and Aristotle say that the natural scientist ought to give this sort of a twofold answer, to explain why it must be, and why it is better so.[371] The explanation of the necessity is based wholly on the material and corporeal (e.g. that

10 15 20 25 30 369,1

5 things must be made up of the hot, the cold, the dry and the wet), while the explanation of the teleology is that by which it is shown that it is better in this way than in that. But 'better' must not be taken simply, but in relation to the nature of each entity. For what is simply and universally best cannot be available to mortal nature; where it is, it is available to eternal and divine nature. In that case the end of all things would be one and the same, since the best is a single thing. 10 Examples of what he means by 'in relation to each nature' are: wings are both necessary and useful for birds, as are feet for land animals, the belly for reptiles and fins for fish. This is the message of Aristotle's treatise *On Parts of Animals*, where he presents his findings under two headings – the necessary and the teleological.

<CHAPTER 8>

15 **198b10** First, we must say why nature is included among final causes. Then we must discuss necessity and explain in what way it is inherent in natural entities. For everybody reduces things to this cause on the grounds that, since hot and cold and suchlike are what they are by nature, then they exist and come-to-be through necessity. For even if they posit another cause, for example friendship and strife or mind, they merely touch on it and then abandon it.

Two questions are raised in this passage. One is to prove that nature is included among causes that act for the sake of something, since he has often touched on this controversial point. The other concerns necessity, to explain in what way it is present in natural entities, 20 since all the early natural scientists reduce causes to necessity, claiming that such-and-such a thing comes-to-be in such-and-such a way through necessity. They reduce things to matter on the grounds that this comprises necessity, saying that a particular thing comes-to-be as a result of that particular quality in the substrate through necessity. For example, it is because heat is light and moves upwards, while cold is heavy and moves downwards, that the universe is as it 25 is, with earth at the bottom and the heavens at the top. For if they posit any other cause, as Empedocles makes Friendship and Strife, and Anaxagoras makes Mind efficient causes, they mention it only so far as to touch on it, but make no further use of it in their explanation of causes. Socrates seems to be criticising Anaxagoras for this in the *Phaedo*.[372] Therefore Aristotle must define material 30 necessity and explain in what way it is present in natural entities, i.e. what is the character of such a cause. Does what comes-to-be do so because of it? Now it cannot do so without it, although it does not do so purely because of it; for this reason it is a necessary cause and an essentially necessary cause. This is what he will explain to us soon.

It is useful in many ways for the completion of the discussion about 35 causes to show that nature creates for the sake of something. For if 370,1 it is especially necessary to know the causes in order to be able to give the reason 'Why?' by referring it to them, and if this is profitable because it is assumed that nature creates for the sake of something and not at random and purposelessly, then the truth of this assumption must be demonstrated. Further, if the form is posited as a final 5 cause and a goal on the grounds that nature creates all that it creates for the sake of this, then it must be demonstrated that nature creates for the sake of something. Furthermore, of the early natural scientists some claimed that luck and chance were the efficient causes of what comes-to-be (or rather that things come-to-be without a cause), while others were satisfied with material necessity.

Having shown, then, that luck and chance are *per accidens* causes, 10 and that what creates something *per se* and for some purpose must be prior to a cause that is *per accidens* and supervenes to no purpose (at least as far as the completion is concerned), he needs to show just what this is. He has in any case demonstrated from his examination of causes that there is an efficient cause of natural entities, and has stated that this is nature, which differs from luck and chance in that it acts as something prior[373] and for the sake of something; he now 15 must show that nature does act for the sake of something, and he must also discuss material necessity. Furthermore, throughout the book in his discussion of causes he has made it a special feature of causes (a) that the efficient cause – nature – does not act at random and purposelessly but for the sake of something, and that the final cause is the form, while (b) so-called material necessity is designated 20 so not on the grounds that what comes-to-be does so because of it (for this belongs to the final cause) but because it could not do so without it. This is what is necessary in the strict sense of the word – taken as by necessity but not primarily so.

198b16 But there is a difficulty: why should we imagine that 25 nature acts for the sake of something and because it is better so? Why does it not act in the way that the rain falls – not in order to increase the harvest yield, but from necessity? For the vapour that is drawn up must become cooled, and when cooled precipitate; when this happens the corn grows *per accidens*. Similarly if someone's crop rots on the threshing floor, the rain did not fall for this purpose – to rot it – but it happened *per accidens*. Why, then, should this not be the case with parts of natural entities? Why should teeth not come through as a result of necessity, the front ones sharp and suitable for biting, the back ones broad and suitable for chewing food – *per accidens* and not through intention? Similarly with other parts where purpose seems to be inherent. So when things all turned out just as they would have done if they had come-to-be for some purpose, they survived because they were suitably constituted by chance; but things not so constituted perished and continue to perish, for example the man-headed calves mentioned by Empedocles.

This and similar arguments might well cause difficulties.

Aristotle proposes to show that nature acts for the sake of something. First, as is normal practice for him and also for his teacher Plato, he gives a good account of the opposite argument, defending the account
30 as if he intends to agree with what its originators claimed. This serves to leave no grounds for contradiction to those who might be able to bring some subtle objection at some point. So he now begins by pleading the case of those who say that nature does not act 'for the sake of something and because it is better so'. For those who want nature to be acting for the sake of something press their claim on the grounds of advantage, especially in the case of parts of animals; for
35 if the front teeth grow sharp in order to bite the food, while the back
371,1 teeth grow broad in order to chew it, it is clear that they grow not at random and purposelessly, but because their creator has an eye to what is advantageous and better. This is their claim, but those who do not want nature to be acting for some purpose say that what comes-to-be does so as a result of natural or material necessity, and
5 that it just happens to be useful or advantageous *per accidens*, since nature is not acting for this purpose; but just as Zeus might send rain during the summer, but not in order to rot the grain on the threshing floor (the rotting of the grain being a *per accidens* outcome of the rain), what is there to prevent us from claiming that it does not rain in order to increase the yield, but that it rains anyway and that the crops grow *per accidens*, and that the rain happens by nature and material
10 necessity? For the vapour is drawn up, cools, condenses and as it grows heavy it falls – this is what rain is. In any case they would never say that either mind or nature would cause what is better to happen for the sake of what is worse.[374] Consequently, the motion of the sun, which causes warmth, rain and different kinds of weather by a process of cooling, does not happen for the sake of the crops;[375] it
15 is rather the case that their growth – like many other occurrences – supervenes *per accidens*, *as if* some very shrewd mind were at work.

So what reason is there why parts of animals and of plants should not be produced as a result of material necessity, without nature aiming at the good and useful or acting for the sake of something?
20 For the front teeth grow sharp and the back teeth grow broad not

because of any advantage, but because of natural necessity. For one could say that the gum at the front of the mouth is more solid than it is on either side at the back, on account of the curvature; for when things are curved they are more solid, since the concave surface is compressed and the convex surface stretched, particularly at the place where the curvature is at it greatest. What passes through something that is more solid becomes sharp, and what is already 25 sharp comes out more so. Since these things occur in this way because of natural necessity, it is without purpose and *per accidens* that they are for the benefit of animals, and it is only *as if* they occurred for the sake of this. For why should some animals be destroyed by their own parts, as eagles are through hunger when their beaks become twisted, while others are preserved, if it were not the outcome of chance? Where everything happens to coincide *as if* happening for the sake of 30 something, even if they are really the outcome of chance, since the coincidence is of benefit, then they survive; otherwise they perish and are destroyed. For example, Empedocles says that at the beginning of the Rule of Love[376] parts of animals, such as heads, hands and feet, started by coming-to-be at random, and then united as 'the offspring 35 of cattle with the heads of men, while others in turn sprang forth as 372,1 the offspring of men' – with the heads of cattle, of course – i.e. hybrids of cattle and men. All those that joined together in such a way as to achieve survival became animals, and their continued existence depended on the fact that they meet each other's needs; the teeth cut and chewed the food, the stomach digested it and the liver turned it 5 into blood. When the head of a man joins with the body of a man it causes the whole to survive; but since it does not match the body of an ox, it perishes. For any parts that do not join together according to the appropriate formula perish. This is universally the case even today.[377] Those ancient natural scientists who said that material 10 necessity was the cause of generation seem to be of this opinion, as do the Epicureans in more recent times. Alexander says: 'Their error lay in the fact that they imagined that anything that occurred for the sake of something did so as a result of choice and reasoning, while failing to see that the products of nature occur in this way too. But they are wrong, as we have already stated; in fact Aristotle says that nature acts for the sake of something but not according to reason.'[378] 15 Perhaps if nature were the first and strictest cause, it ought to act foreseeing that for the sake of which it is acting; but when it is in the service of some agent which is acting in the strict sense and for the sake of something, it is itself too acting for the sake of something; but it does not take previous consideration. For example, the axe cuts for the sake of something; it does not take previous consideration, but it is cutting in the service of someone who is doing so. 20

198b34 It is not possible, however, that this should be the case. For these and all natural phenomena occur as they do either always or usually, while the results of luck and chance do not. It is not considered to be the result of luck or coincidence that it often rains during the winter, but it is if it rains in midsummer; similarly it is not considered to be the result of luck or coincidence if there is a drought in midsummer, but only during the winter. Therefore if we think that things occur *either* as the result of some coincidence *or* for some purpose, and if such things do not occur as the result of coincidence or chance, then they must be for some purpose. But all such occurrences happen by nature, as admitted by those who put forward these arguments; therefore there is purpose in whatever is or comes-to-be by nature.

He has supported the argument which states that for things that occur by nature necessary and advantageous concomitants are *per accidens*, and that nature does not act for the sake of need and
25 advantage. He now shows that it is not possible for this to be the case, taking as evident the conclusions that he later draws, (a) that things happen *either* in relation to some predetermined goal and for some purpose, *or* as the result of luck or chance, and (b) that the parts of animals discussed, and in general all natural phenomena, occur as they do always (as is the case with things eternal) or usually, while the results of luck and chance are shown to belong among what happens less often. The rest of his argument is as follows. Natural
30 phenomena occur as they do either always or usually (for a man always begets a man, as does a horse a horse), and the nature of the teeth and of the other parts is almost always constant, while the results of luck and chance (in general any *per accidens* occurrence) do
373,1 not occur as they do always or even usually (such occurrences were shown to belong to what is not usually the case). Therefore natural phenomena are not the result of luck or chance, and are not generally *per accidens*. The syllogism is drawn in the second figure,[379] where Aristotle has set out its premisses, but has not explicitly drawn the conclusion. He demonstrates the major premiss which states 'the
5 results of luck and chance are not as they are either always or even usually' from the converse; for we do not say that what usually happens (e.g. frequent rain in winter or drought in summer) is the result of luck or coincidence, but only when it rains frequently in summer or a drought occurs in winter – these in general belong to what is not usually the case. It is possible to draw the argument
10 according to the first figure as follows.[380] What occurs by nature does so usually; nothing that occurs usually does so *per accidens*; therefore no natural occurrence is *per accidens*.

He has therefore shown that for natural phenomena necessary and advantageous concomitants are not *per accidens*, since they occur as they do usually; he now proceeds to show that they are for some purpose. The disjunction which he applies is as follows. Whatever 15 happens does so *either per accidens* and as a result of coincidence *or* for some purpose; but the occurrences under discussion are not the result of coincidence, as has been proved; therefore they are for some purpose. The disjunction seems to be a necessary one, since even his opponents talked of coincidence as excluding purpose. For since in every case what occurs must do so either in the one way or the other, 20 they thought to do away with purpose by demonstrating that <things happen> *per accidens*. In this way, if anyone begins by understanding that whatever occurs by nature does so always or usually, he draws the conclusion (from the premisses and from the disjunction) that natural phenomena occur for some purpose, both in the second and third figures.[381]

Aristotle himself continues the argument in the case of the parts of animals and suchlike on the grounds that it is evident that these 25 occur as they do usually. He shows that the efficacy in the case of the parts is for some purpose precisely because it is not *per accidens*, and he adds: 'all such occurrences happen by nature, as admitted by those who put forward these arguments', since animals, plants and their parts, in which there is nothing superfluous or deficient, are the works of nature. He has made the parts of animals and suchlike his subject matter, and has affirmed purpose and the activity of nature 30 to be the case with them. He draws a particular conclusion according to the third figure, that some of the things that occur for a purpose occur by nature. For Alexander shrewdly observes: 'We should look not in the first figure but in the third for the reasoning which leads to the particular conclusion which states that some of the things that occur for a purpose occur by nature. For the conclusion according to the first figure states that all things that are for a purpose are 35 products of nature, which is not true. For although all natural phenomena are for a purpose, not all things that are for a purpose are products of nature. For the products of choice and art are among 374,1 the things that are for some purpose. The conclusion, then, is as follows. Since animals, plants, their parts and their efficacy almost always occur as they do, and since what almost always occurs as it does does so for some purpose, if we add the other premiss which states that animals, plants and their parts occur as they do by nature 5 (which is agreed by his opponents), then it will be concluded that some of the things that occur for a purpose occur by nature. This fact he has himself too demonstrated in the words "therefore there is purpose in whatever is or comes-to-be by nature". For if there is purpose in whatever is or comes-to-be by nature, then everything that is a 10 product of nature is for some purpose, but not *vice versa*. For example, since nourishment and growth are found in animals, animals are nourished and grow – but not everything that is nourished or grows is an animal.'

Perhaps Aristotle was dealing with the subject in a concise and general manner when he said that 'all natural phenomena occur as they do either always or usually', adding that 'the results of luck and
15 chance do not', drawing his conclusion according to the second and the first figures, as has been said above. Yet when he continued the argument by what is evident in the case of the parts, he drew a particular conclusion according to the third figure. So if the growth of the crops because of rain is natural and usual, it is clear that their growth because of the rain is not *per accidens*; therefore the rain falls
20 for a purpose, namely the growth of the crops,[382] just as it falls for a variety of other purposes which are not consequential on the rain *per accidens*; for the same event can occur for a variety of purposes, both major and minor.

Does it then rain in order to rot the crop? No, for this is not one of the usual results consequential on the rain, but one of the less usual ones. Therefore even if it occurs by nature, it represents a failure on
25 the part of nature to achieve its goals – just as freaks are. Unless, of course, the rain falls in order to rot the crops; but in that case it is not nature that causes the rain to fall for that purpose, but the creator god, who ordains all things according to justice, exploiting nature.

199a8 Furthermore, where there is an end in view, the earlier and successive stages of the process are done for the sake of that end. Therefore each thing is accomplished according to its nature, and its nature accords with its accomplishment, if nothing prevents it. Things are accomplished for some purpose, and so their nature accords with that purpose. For example, if a house were one of the things produced by nature, it would be produced in just the same way as it now is by art; and if the products of nature came-to-be not only by nature but also by art, they would be produced in just the same way as they are now produced by nature. Therefore the one is for the sake of the
30 other.

This is a second proof. Alexander explains the passage as follows: 'He has shown that the products of nature are for some purpose (for they possess an end "for the sake of which"). He adds the following statement as a consequence, that where there is an end in view, the previous stages in the process are done for the sake of it; it follows
375,1 from this that the previous stages happen for the sake of the end even in the products of nature.' However, I do not think that I grasp his

point, since he seems to me to be saying nothing more than that, in understanding that the products of nature are for some purpose, he concludes that the products of nature are for some purpose. For what do the words 'the previous stages happen for the sake of the end even in the products of nature' mean other than that? Alexander also says: 'Furthermore, in taking as agreed the fact that where there is an end in view the previous stages are done for the sake of it, he adds the statement that each thing is accomplished in accordance with its nature; from this he concludes that in the products of nature there is not only an end in view, but also that it is their nature to be for some end. He adds this proof in case anyone should claim that, although there is an end in view in the products of nature, even so nature does not produce them for the sake of that end.' Yet it does not make sense to say that the very agent of production produces what it produces, but not in order to achieve the end in view. Generally speaking, if it is taken as agreed that, where there is an end in view, the previous stages in the process are done for the sake of the end, it is also assumed that they are done for this end by the agent.

Perhaps, then, the first proof is as follows: in the case of things that come-to-be, where there is an end in view in the form of a termination at which the change finishes and the forward process comes to a halt, in their case all the stages before the termination are for the sake of it; there is an end in view in the case of the products of nature, since the change is continuous and proceeds without interruption until it secures the achievement of the end. (Although the blade, the stalk and the ear are complete entities in themselves, they are in fact stages on the way to the complete plant, and it is only then that there is a cessation and completion of the process.) The conclusion which follows from these two premisses is that, in the case of the products of nature, the earlier and successive stages of the process are always done for the sake of the end; in other words, nature acts for some purpose, which is what the proposition was. But of the two premisses Aristotle presented only the former, saying: 'where there is an end in view, the earlier and successive stages of the process are done for the sake of that end.' He left unsaid the second premiss and the conclusion as evident.

It is clear that both premisses are true; for where there is some end in view, and a continuous and established process towards it, it is obvious that all the earlier stages are for the sake of that end. For example, in the field of human intention health is an end, and the doctor goes through all the earlier stages for the sake of the health of his patient. It is obvious too that in the case of natural entities the <perfected> form is the end, and that everything takes place in order that each natural entity should achieve that perfection and attain its proper form. Therefore the activity of nature proceeds towards it, and

35 halts when it reaches it. If these points were established, then it would be true that in the case of natural entities all the earlier stages are for the sake of the end. Therefore, if this is so, the end does not 376,1 merely supervene on the earlier stages, but they take place according to their nature for the sake of the end, and 'each thing is accomplished according to its nature, and its nature accords with its accomplishment, if nothing prevents it'. So if it is accomplished for the sake of the end, then it exists for the sake of the end according to its nature. Perhaps the words 'Therefore each thing is accomplished according 5 to its nature etc.' show that the end is nothing supervenient. For if Aristotle takes as agreed the fact that each thing is done for some purpose, and adduces as a consequence that its nature accords with the purpose, and if what is done for the sake of some end does not have that end as something supervenient, then it is clear that there is no gain in changing 'what is done for some purpose' to 'is for some purpose according to nature'.[383]

10 Alexander, as has been stated, says that the additional point about nature was made 'in case anyone should claim that, although there is an end in view in the products of nature (as was the case with the teeth), even so nature does not produce them for the sake of that end, but the necessity supervened'. He goes on to say, as he will remind us as he proceeds: 'Furthermore, even in what occurs as the result of luck there is a purpose and its intended achievement; but the earlier 15 stages are not done for the sake of that purpose. Therefore Aristotle, having shown that the products of nature are for some purpose, now adds the point that he does not mean "for some purpose" in the way that purpose was involved in the outcome of luck, but because it was according to their nature.'

But perhaps Aristotle did not add the words 'according to their nature' to make a distinction between them and the results of luck. For in the case of the results of luck what is done is done to no purpose 20 only with regard to the end in view. As it is, he takes as agreed that the process is for some purpose, and it is on this basis that he brings in the phrase 'according to their nature' in such a way that not even Alexander's first explanation seems to hold good. For if it is taken as agreed that whatever is done in the case of natural entities is done for some purpose, nobody would say that there is an end in view in the products of nature, but that nature does not act for the sake of that end. Instead, perhaps, the point was made in contrast to the 25 products of art, since in their case what is done is done for some end, but not according to nature.

Aristotle has therefore shown in general that 'where there is an end in view, the earlier and successive stages of the process are done for the sake of that end', which is consistent in the case of the products of both art and choice. He now shows that in the case of natural entities this is so 'according to their nature'. Nature is that which acts 30 for some purpose; for this is what he means by the phrase 'according to their nature'. I judge that this is his meaning from the words that follow: 'For example, if a house were one of the things produced by nature etc.', by which he indicates nothing other than that just as in the case of the products of art what is done is done for some purpose by *art*, i.e. the art or the artificer is acting according to art, so in the 35 case of the products of nature what is done is done by nature for some 377,1 purpose. For if 'a house were one of the things produced by nature', it is in this sense that the earlier and successive stages of the process of housebuilding would be done by nature for the sake of the end in view – just as they are now done by art and in no other way. Even if art were fashioning an animal, it would not fashion it in any way other than that in which nature fashions it, and it would have 5 completed the previous stages for the sake of the end. That is what is meant by 'the one is for the sake of the other'. He introduced this idea of reciprocity, showing that even if one of the products of nature were produced by art as well, it would still be produced in the same way. His earlier example of the house was a good one, since homes are made by nature by irrational animals, and by art by humans. So 10 if a house were to be produced by art just as it would be by nature, and if in the products of art the stages prior to the achievement of the end are obviously done for the sake of the end, then this is also the case in the products of nature.

When he says: 'Where there is an end in view, the earlier and successive stages of the process are done for the sake of that end', by 15 'end' he does not mean simply 'that for the sake of which' (for that was the question at issue, whether nature acts for the sake of something), but what brings to completion the earlier and successive stages – in general the ordered process – and what perfects the process of change, coming as the attainment of the process which is continuous, and terminating the process of change. Yet Alexander says: 'By "end" he does not mean simply that which occurs later, since 20 even in the case of what happens by chance something happens later, but he means the end in view.' He concludes that 'This is what he now wants to demonstrate, and so he adds the words "Furthermore, where there is an end, that which is for the sake of something ..."; for he is not satisfied simply to say "end", but adds "that which is for the sake of something" to show what sort of end he has in mind.'[384] The following points seem worth noting: first, that the passage in all the 25 manuscripts that I have read does not read like that, but has: 'Furthermore, where there is an end in view, the earlier <and successive> stages of the process are done for the sake of that end'; and secondly that it is not 'that which is for the sake of something', but 'that for the sake of which' which is the end. Wanting to construe it

as 'Furthermore, where "that which is for the sake of something is an end",' I find that the grammar is inconsistent on account of the singular definite article that goes with 'for the sake of something'. In
30 general, as I said, the question at issue was whether nature acts for the sake of something. This he demonstrates from the fact that nature makes its way in succession and in an ordered manner until it arrives at a state of completion; such a state is 'that for the sake of which'. What proceeds in this way proceeds by nature, as he said earlier and as he will proceed to show when he says: 'A product of nature is anything which moves continuously from a principle within itself,
35 and arrives at some end.'

378,1 **199a15** In general art either completes what nature cannot bring to perfection, or else it imitates. Therefore if the products of art are for some purpose, clearly those of nature are too. For in both cases the relationship between cause and effect is the same.

On the basis that even they[385] agree that art acts for some purpose, Aristotle has shown that nature too acts for some purpose; because
5 he has made his own use of the demonstration he now confirms it by showing the similarity between art and nature, and the common ground between them. The common ground he demonstrates from the fact that art works together with nature to complete what nature cannot effect by itself; for example, medicine brings about health and the regaining of lost weight when nature is unable to do so by itself. Art imitates nature by creating two- and three-dimensional repre-
10 sentations of animals and plants. These days some people can hatch chicks from eggs in the absence of the parent birds by simulating them, through art. So if art works together with nature or imitates it, and if the products of art are for some purpose, it is clear that the products of nature are too. For the end stands in the same relationship to the means in the field of art as in the field of nature, since art
15 imitates nature and shares common ground with it. So if in the field of art the cause is for the sake of the effect, it is similarly so in the field of nature. Yet Porphyry says that in this case the reasoning is *a fortiori*: 'For if art, which is inferior to nature as image is inferior to model, effects the means for the sake of the end, all the more will
20 nature, in that it is superior to art, act in the same way.' But in what sense could he claim that art is inferior to nature, since Aristotle says that art completes what nature cannot bring to perfection? Unless, of course, he meant it in the sense that art makes good the deficiencies (even what is inferior can achieve that) or else produces imitations.

199a20 This is most obvious in the case of other animals, which do not act by art or investigation or planning. As a result some people wonder whether it is through intellect or something else that spiders, ants and suchlike achieve what they do. At a lower level it is clear that even in the plant kingdom things happen which contribute towards an end, for example, leaves grow in order to protect the fruit. Therefore if it is by nature and for some purpose that swallows make their nests and spiders their webs, and if plants produce their leaves for the sake of the fruit, and send their roots down rather than up in order to gain nourishment, then it is obvious that such causation is at work in what is produced and exists in nature. 25

He argues very systematically from the realm of irrational animals to show that what is produced in nature is no less for some purpose than the product of art, choice or intellect. For spiders and ants do not act by art or choice, but irrationally and according to their nature; 30 but they are clearly acting for some purpose, to the extent that it can be argued that they are perhaps acting with intellect,[386] art and forethought. They do everything to meet some need or advantage, as 379,1 do those which employ intellect or forethought; in order to catch flies spiders weave webs which are sturdy and spread over a considerable area, while ants store up their food, are marvellously systematic in dragging things many times their own weight, often bring their food, if it has become damp, into the sunshine to dry it, and march in 5 orderly columns. What is even more miraculous than all this is the fact that they divide their nests, which we call anthills, into three parts; in one they live, in another they store their food and in the third they bury their dead. The swallow and the nightingale build their nests by selecting the most malleable clay (so that even doctors use 10 it instead of desiccating drugs) and by binding the clay with straw so as to provide the strongest and most spacious unit. The bee too excites wonder because of the arrangement of its hexagonal cells. The kingfisher's nest, although built under water, is in no way damaged by the water. One could find much to say to demonstrate a quality of art in the creations of irrational animals. But in case anyone should think 15 that irrational animals act because of any power of reasoning (*logismos*), he should descend to the level below rational and irrational animals, and consider purpose in the plant kingdom, where there is no room for doubt as to whether they perhaps employ intellect and art. He will see that even here nature acts for some purpose. For the leaves and the husks grow for the sake of the fruit, and the roots for the sake of nourishment. For why should the roots not grow upwards 20 and the branches downwards? Why do the plants protect the seeds so securely unless they have real foresight for mutual regeneration?

The conclusion of the argument might be as follows: what is produced by nature in the realm of irrational animals, and more particularly in the realm of plants, is produced in the same manner 25 as what is produced by intellect and art; what is produced by intellect and art is for some purpose; therefore what is produced in the realm of irrational animals and the realm of plants is produced for some purpose; therefore nature acts for some purpose.

I think it worth observing at this point that Aristotle assumes that irrational animals and plants are natural things, although they are inhabited and organised by souls,[387] and irrational animals enjoy 30 imagination, perception and appetite in their own activities. If he had wanted to see purpose in things that are strictly and solely by nature,[388] he would have adduced the local movement of the elements as occurring in order to achieve their proper unity,[389] and the warring of the opposite qualities as occurring for the sake of their preservation in the substrate, and the earlier of the changes that happen in the creation of each thing as always occurring for the sake of the sub- 35 sequent one. For fire comes into being from water through the 380,1 mediation of vapour and air. So perhaps Aristotle gives the name 'nature' to everything in the soul that is to do with bodies. That is why people consider that *On the Soul*, which deals for the most part with soul of this kind, to be part of his researches in natural science.

5 **199a30** Since nature is twofold, both as matter and as form, and since the latter is the end, and since everything else exists for the sake of that end, then form would be the final cause.

He has shown that the products of nature are for some purpose. He now wants to demonstrate what this purpose is. Therefore since natural entities are compounds of matter and form, and since both of these are called 'nature', as has been proved above, and since form 10 deserves the title rather than matter, and since in the case of something that is being produced the form is the end, and since every stage before the end is for the sake of the end, then the form would be that for the sake of which everything else happens, as he showed earlier, when he made the identification between form and end. Alexander says: 'It seems to me that from this he proves that the cause is more properly the form than the matter, as the ancients seemed to say when they reduced everything to the material cause. 15 For if the matter exists for the sake of the form, then the form would be more properly be the cause than the matter.'

199a33 Even in the products of art mistakes can occur. For writers make errors in their writing,[390] and doctors in their prescribing. So clearly this can happen in the products of nature too. Indeed there are occasions in the field of art when a correct procedure is directed to some end but when the attempt to achieve that end goes wrong; it could be the same among the products of nature, and deformities[391] could be seen as failures to achieve the end. So in the original combinations if the young of cows were unable to arrive at some standard and some goal, they would have been born as the result of some corrupted principle, just as these days happens in the case of deficient seed. Furthermore, the animal must not come immediately, but the seed must come first. The 'whole-natured shape that came first' was the seed.[392]

An objection can be made, against those who claim that nature acts for some purpose, from the existence of monsters. For what would be the purpose in the birth of what Empedocles calls 'the man-faced young of cows'?[393] Aristotle now meets that objection too by saying that not everything that is produced for some purpose necessarily succeeds, but that there are often failures. In fact there are failures among the products of art, although it is universally agreed that they are produced for some purpose. Such art was not operating for the sake of the failure, but for the sake of the end that was not achieved. In this way nature too acts for some purpose, but often that purpose is not achieved and something other than it results. Therefore just as medicine, even if it does not achieve the end aimed at by the doctor who administers it, is nevertheless administered for some purpose; and just as the writer, even if he makes an error, nevertheless writes in order to write correctly; in just the same way nature, even if it fails to produce a human being, still acts in order to produce one – the failure is of just the same kind as with the products of art. So monsters too represent failures to achieve a purpose. That is why, I imagine, they do not fall far short of achieving that purpose. A human is usually born from a human seed; if that is not so, it is an animal, and never a fig- or plane-tree. Such are Empedocles' 'man-faced young of cows'. When Aristotle says that even among the products of nature deformities occur he means that they are 'failures on the part of what is acting for some purpose'. When he says that such deformities occur in the same way as failures among the products of human design, he is indicating that Empedocles' 'man-faced young of cows' or 'ox-faced young of humans' (if such things did occur) are born because of a failure in the principles which are the origin of all created and material things – just as these days monsters are the result of a failure in the seed, when it is feeble and 'the principle is corrupted'. Such examples are not sufficient to prove that nature does not act for some purpose.

20 25 30 381,1 5 10

To begin with he agreed with Empedocles when he said that the 'man-faced young of cows', monstrous animals, are from a principle; this being granted, he met the objection on the grounds that even
15 these monsters were born 'as the result of a corrupted principle'; he then refuted the grounds for such an objection, which was based on Empedocles' monsters, by showing that the birth of monstrous animals from a principle was untenable; for if he actually said this, and if monstrous animals were born from a principle, the fact that nature acts for some purpose is not refuted. For nature does act for some purpose, and if a mistake occurs in either the agent or the material
20 it produces some other outcome and not the one which the agent was acting to achieve. Moreover, the grounds of the objection are false. For it is not possible for animals to be born from some principle, but the seed must precede the animals, since it is necessary for all animals, both perfect animals and monsters, to be born from their appropriate seed and not instantaneously, with half animals at-
25 tached to half animals. If it is necessary that each animal should get the principle of its birth from its appropriate seed, it is clear that there is some order in their birth – a first stage and a second stage – and that the first stage is for the sake of the second. The idea of purpose lies wholly within such considerations.

30 In the second book of his *Physics*, before the passage on the assembling of male and female bodies, Empedocles says:

> Come now, hear how fire, as it was separated out, raised up the nocturnal shoots of men and women full of lamentation; for it is no misguided or ignorant account. First, whole-natured shapes sprang up from the earth, having a share of both water and
> 35 shape;[394] these fire emitted, wishing to meet its like, although
> 382,1 they did not yet display either the desirable form of limbs nor voices nor such things as are the parts particular to males.

In the light of these words of Empedocles Aristotle points out that Empedocles himself appears to mean that the seed is produced before the animals. The phrase spoken by Empedocles, 'the whole-natured
5 shape that came first', was the seed, which did not yet display 'the desirable form of limbs' because it was liquid in actuality, being fluid, but the form of man in potentiality. But perhaps it was impossible for either the animal to exist before the seed, or the seed before the animal. For just as each plant and animal is born from its own
10 particular seed, so each seed is produced by its own particular animal or plant. Aristotle might have pointed out that Empedocles does not say that the seed *always* exists before the animal. And it is clear that the animal does exist before the seed. He confirms his observation by concluding that ' "the whole-natured shape that came first" was the

seed', since even Empedocles himself realised that the seed must exist before the birth of the animal. If it is the seed, then it seems to me remarkably suitable to attribute the term 'whole-natured' to it. For strictly speaking it is whole-natured because it is that which is, throughout itself, entirely what it is if no differentiation has yet taken place within it. For every part of the seed is all the parts of the body, but no part of the body is the other parts once the differentiation has taken place in them and the whole nature has been dispersed. 15 20

199b9 Furthermore, purpose is to be seen in plants too, although it is less clearly articulated. So in the plant kingdom were there or not 'olive-faced offshoots of vines', just as there were 'man-faced young of cows'? It would be absurd if it were the case, but it ought to be so, since it is true of animals. Also in the case of seeds anything whatever would have to be produced.[395]

This is said to counter the objection that arises from what Empedocles says. Since purpose is evident even in the plant kingdom, as was stated earlier, with the leaves growing for the sake of the husk, and the roots growing for the sake of drawing nourishment, are there here too, just as in the animal kingdom, some deformities – 'olive-faced offshoots of vines' from the vine and the olive – so that we can, on the basis of this object that nature does not act for some purpose? Or is this not the case? For if it is, then it was absurd of Empedocles not to say so. But if it is not the case, then how can it be true of animals but not of plants? Alexander says: 'It seems to me that by adding "although it is less <clearly articulated>" to "furthermore, purpose is to be seen in plants too" he is implying some *a fortiori* argument like the following: if in the case of those products of nature where the purpose is less obvious these products do not grow at random but for some purpose; and if in the case of those plants where the purpose is less clearly articulated what grows does not grow by chance – not even "olive-faced offshoots of vines" – then not even among animals, where purpose is more clearly articulated, do chance animals emerge, so that "man-faced young of cows" do not emerge.' Aristotle adds 'Also in the case of seeds anything whatever would have to be produced', <which would be so> if in fact natural entities occurred by accident and chance. For example, from a particular seed on different occasions a horse, a man or some other animal would have to be born. As it is, it is entirely determined what will be produced by a particular seed – man from human seed, horse from equine seed. Therefore the prior stage happens not by chance and at random, but according to order and sequence and for the sake of the subsequent stage. 25 30 383,1 5 10

199b14 In general a person arguing in this way does away with nature and the products of nature. For the products of nature are those things which develop continuously from some source within themselves and arrive at some end. What develops from each source is not the same in each case, but it is not a chance outcome; but the development is always to the same end, unless something prevents it.

15 Just as he said earlier that those who deny change are not natural scientists, because they deny nature itself, since it is established as the source of change and its cessation, so now he shows that those who deny that nature acts for some purpose deny natural entities and nature itself. But they fail, in their observation of such entities, to explain how they happen‡[396] to occur. Aristotle demonstrates this 20 from the notion that we hold about nature and its products. For if nature is a principle of change and its cessation in that in which it resides, and if the products of nature are things 'which develop continuously from some principle inherent in them and arrive at some end', then not every principle produces the same end (for the development from the human principle, such as the human seed, does not 25 result in a dog, which is what the development from a canine seed results in), but the development from each specific principle or process is always to the same end, the appropriate one, if nothing prevents it (for the human develops from the human principle, and the horse from the equine principle); consequently, it is not the same thing that develops from the totality of different principles, nor does something random‡ grow from each of them. If the case is such that each thing develops continuously and arrives at some determined 30 end, and the subsequent stage always follows on the previous one, then it is clear that in such cases the previous stage always happens for the sake of the subsequent one. So if the products of nature are for some purpose, what is not for some purpose is not a product of 384,1 nature. In general if we do not hold such a notion about the products of nature – that they occur for some purpose – we are not even thinking about what is a product of nature. For if we deny development to a determined end we deny nature itself. For it is a principle of that sort of a development, and not of random‡ change. Alexander lays out the proof as the following syllogism: 'The products of nature develop from some principle within themselves and arrive at some 5 end;[397] what does not occur for some purpose does not develop continuously from some principle within itself and arrive at some end; therefore what does not occur for some purpose is not a product of nature; consequently, those who say that the products of nature are not for some purpose but occur by chance, these people deny nature and its products.'

199b18 An end and the means to it can occur by luck. For example, we say that it was by luck that the stranger came, paid someone's ransom and then left, when he only did it *as if* he had come to do it, but in fact did not come for this purpose. This outcome is *per accidens* – for luck is among the *per accidens* causes, as we said earlier – but when this happens always or usually, it is not *per accidens* or the outcome of luck. Natural entities are always like this, if nothing prevents it. 10

Since both the end and the means to the end seem to be outcomes of luck when we say that 'it was by luck that the stranger came and paid someone's ransom and then left, when he only did it *as if* he had come to do it', even if in fact this was not his intention (for this is a *per accidens* cause), it is for this reason and in this case that both the end and the means are both said and thought to be outcomes of luck. An example is to be found in Menander's play when Demeas pays Crateia's ransom. (Some <Aristotle> texts read, instead of 'and then left', 'and then released him' – both readings are found, just as we find either 'paid someone's ransom' or 'washed someone'.[398]) But there is a difference between this case and a strict application of the principle in cases that pertain to nature. For what happens 'always or usually' is not *per accidens* or the outcome of chance – which happens less often; but 'natural entities are always like this, if nothing prevents it'. 15 20

Consequently, the end and the means to it pertaining to the outcomes of luck are different from those pertaining to natural entities, because the former are *per accidens* and less often, while the latter are usually the case.

But in what general sense can we talk about ends and means in the above example, since the man did not come with the end of paying a ransom in view? The answer is that on the one hand the end would be an outcome of luck because the end – the paying of the ransom – followed on his arrival as something advantageous, even if he did not come for that purpose (this is an end in a sense, if not strictly so); on the other hand the means – the arrival of the stranger – appear to be directed towards this end, even if in fact he did not come in order to pay a ransom. So it is in this sense that Aristotle's argument is admissible. Now even if, conversely, the stranger did not come with this end in view, nevertheless luck allowed his arrival to happen for this purpose and produced this outcome to his arrival. That is why Aristotle agreed that ends and means apply to such a case, even if such an outcome was not consistent with the original choice. He showed the difference between the two cases by applying the terms 'more often' and 'less often'. 25 30 385,1 5

199b26 It would be ridiculous for anyone to imagine that something has not happened for some purpose unless he can see that the agent of change has deliberated. Yet art does not deliberate; if ship-design were inherent in the timber, it would act in just the same way as nature. If, then, purpose is inherent in art, it is also inherent in nature. This is most evident when someone looks after his own health; nature is like that.

It is clear therefore that nature is a cause and that it is directed to some end in this manner.

10 He has given his subject matter a varied treatment; he now corrects the thinking that leads some people to the conclusion that nature does not act for some purpose. For they imagine that anything that acts for some purpose makes prior deliberation and predetermines the end, and then acts aiming to achieve it; they get this notion from what occurs in the field of art. So Aristotle's reply is based on that very field which occasioned their distorted reasoning. He says that 'art does not 15 deliberate'; for craftsmen do not need deliberation in order to realise their designs, since each product is determined when it is realised according to the design. Why should a clerk need to deliberate in order to write Dion's name as he does? It is only in the areas of art that require calculation that there is any scope for deliberation.

20 The conclusion of the argument would be as follows: if art acts for some purpose but does not deliberate beforehand, *a fortiori* there is no need for nature to deliberate beforehand when it acts for some purpose. The former statement is true, so therefore is the latter.

He further demonstrates the similarity between nature and art in his words 'if ship-design were inherent in the timber, it would act in just the same way' as if nature were building the ship; for the products 25 of nature differ from the products of art only in the fact that the former are initiated from inside themselves and the latter from outside. That is why he proceeds to an even closer parallel, saying that the similarity between art and nature is 'most evident when someone looks after his own health; nature is like that' – since the cause is internal. But 30 the distinction between such a case and nature has already been stated by him, in that in the one case the principle of change is *per accidens* (since he is not being treated by himself *qua* doctor, but in view of the fact that he is *per accidens* ill), while in the case of the products of nature the cause is *per se*; in so far as they are the products 386,1 of nature, the source of change is inherent. However, the similarity between art and nature is particularly to be found in those cases where the change seems in some way to originate from inside.

<CHAPTER 9>

199b34 Is the term 'necessity' to be applied conditionally or unconditionally? These days people think that necessity is present in the production of something, as if they imagined that a wall had been produced by necessity because heavy material naturally sinks to the bottom while light material finds its way to the top – which is why the stones and the foundations end up at the bottom, while the bricks are above them because they are lighter, with the timber, being the lightest, at the very top.

The wall could not in fact have been constructed without these materials; but it was not constructed because of them (except in the sense that they comprise the matter) but in order to protect and preserve certain things. This is true in all other cases where there is some purpose; things cannot happen without whatever has the required material nature, yet they do not happen because of this – except in the material sense – but for some purpose. For example, 'Why is a saw as it is?' In order that a certain outcome may occur – i.e. for that purpose. But this purpose could not be achieved unless the saw is made of iron; therefore it must be made of iron *if* it is to be a saw and *if* it is to do its job. Therefore the necessity is conditional, but not as an end; for the necessity is in the matter, but the purpose is in the full account.

After completing his discussion[399] of causes he presents two questions relevant to the discussion of causes. For he makes the distinction which some fail to make between the efficient and the material; he first says that nature is among the causes that act for some purpose and not randomly and haphazardly (which was the view of some); and secondly he explains in what sense so-called material necessity is inherent in natural entities. Certain consequences follow of necessity because of the material, and because of this[400] an advantageous end occurs, as some people seemed to think, assuming that the matter is the strictest and only cause, as the fact of their ascribing the cause of all things to matter shows; is it, then, the case that the matter is inherent of necessity in this way in natural entities, which Aristotle calls unconditional, or is it rather the case that the material necessity follows on conditionally from the appointed end?[401] So if a house is going to be built, i.e. some protection for what is inside it, there must be a roof, with walls and foundations laid below it. Yet even if the material is by nature constituted in such a way that what is heavy sinks to the bottom and what is light rises to the top, the house does not instantly happen just because of this.

20 These, then, are the questions at issue; he rejects unconditional necessity in the sense that those who explain causation in material terms seem to mean: since what is warm rises and is light, while what is moist is heavy and sinks, this is the reason why the universe is constituted as it is with earth at the bottom and fire uppermost, or else because earth is forced to the centre in the cosmic rotation. The 25 former is the opinion of Empedocles, the latter that of Anaxagoras; they ascribe no place to benefit or purpose, but only to the matter, thinking that the end and the advantage were a necessary result of the matter. One might think that the wall too had come into being of necessity in this way, because what is heavy naturally goes to the bottom while what is light naturally goes to the top. Therefore Aristotle rejects this so-called material necessity and accepts condi- 30 tional necessity. He uses it to show that things are not produced on the one hand regardless of their matter, but that on the other hand things are not what they are because of their matter as some sort of overriding cause, but only as their matter and their material cause. For whatever depends on the prior conditions is posterior to them and less of a cause, while the true cause is the end and the purpose. The 387,1 cause in the strict sense of the house or the wall is not what is heavy or light, but protecting and preserving the contents, or keeping the roof up. The matter is conditionally necessary; when the end and the purpose are fixed the material necessity follows, and it is accordingly 5 this which is a *sine qua non* of the completion. For example, if there is to be a house, there must of necessity be stone and timber; yet even if these are available, it is not just for this reason that there must of necessity be a house. So in the case of the products of nature too this sort of necessity must not be ascribed to the end but to the matter,[402] not because the necessity is for the sake of the matter, nor because advantage of necessity is consequent to it, but because it could not 10 occur without it. The stones, bricks and timber cannot motivate themselves, just as they are, to produce a house, but a particular combination, shape and design do everything to perfect the end – though not regardless of the matter. For in fact the matter is one of the *per se* causes, as has been said earlier, but the form, the definition and the essence are causes in the stricter sense. The end does not 15 follow of necessity on the changes in the matter; it is rather the case that <the matter> is by nature preordained to fit in with the end. It is not the case that because there is iron, there must for that reason necessarily be a saw; but when a saw is predetermined, iron must necessarily be taken to make it, so that the matter in this way is subject to necessity not *per se* but because of the form. For there is no necessary consequence of the existence of the matter, while it itself 20 must necessarily pre-exist the form. So the end does not follow its antecedents through any material necessity, but when the end is

established as a condition, then the material necessity follows because of this condition, since what is necessitated by the end, according to which it is necessary that the end follows the antecedents if nothing prevents it, is the end in itself, for the sake of which the antecedents are as they are and acquire the order that they have. Thus the matter is subject to necessity because of the end, but not *vice versa*. For what happens does so for the sake of the end, although not regardless of the matter; however it happens not because the matter is the overriding cause, but because it could not happen without it. 25

The words 'In order that a certain outcome may occur – i.e. for that purpose' indicate the form and the end of the saw. Consequently, those who posit only material causes and claim that things are the outcome of the natural changes in matter would claim that things are the outcome of unconditional necessity. But since form has been demonstrated to us as another cause, in a stricter sense than matter, and since it has been shown that the matter arrives for the sake of the form and that the matter does not introduce the form, not even the matter could be said to be unconditionally necessary, but only conditionally. For the material necessity will only be discovered by laying down the formal conditions. There is, then, necessity inherent in the matter; but that for the sake of which the necessity exists is inherent in the form, where we will find the end and what is strictly the cause. If we manage this, we could forget everything else, and in these circumstances our actions will have an end, which is most fitting for a principle and a cause. In general the purpose is the starting point, and the end is something like this. So Aristotle himself grants necessity to matter, but not in the sense that the form results from it of necessity, but in the sense that the matter can be described as the *sine qua non*. For the end is not of necessity, but the necessity exists because of the end. 30 388,1 5 10

Aristotle is clearly conducting himself in a Platonic manner in this chapter. For Plato says in the *Timaeus*:[403] 'All these things, then, are constituted by nature in this way of necessity, and the creator of what is most beautiful and best in the World of Becoming took them over when he was generating the self-sufficient and perfect god; he used the causes inherent in them as his servants, but he himself fashioned what was good in all that he was creating. That is why we must distinguish two types of cause – the necessary and the divine; we must look for the divine in everything in order to achieve a happy life, as far as our nature allows us to, and we must look for the necessary for the sake of the divine, realising that it is not possible to grasp or apprehend or share in the divine, which is the object of our pursuit, without the necessary.' He makes the same clear distinction concerning these two types of cause in the *Phaedo*[404] when he says: 'Fancy 15 20

not being able to make the distinction, that in truth the cause is one thing, and that without which the cause would not be a cause is something else! Most people seem to be groping in the dark when they
25 call this latter a cause, giving it quite the wrong name. It causes one man to surround the earth with a vortex[405] and cause it to be kept in its place by the heavens, while another supports the air like a board on a flat kneading-trough.[406] But they neither seek the power which causes them to be now so constituted as is best for them, nor do they think that it has any divine force.' Before this passage Socrates has
30 drawn some other lengthy conclusions.

Aristotle has said that 'in all other cases[407] there is some purpose; things cannot happen without whatever has the required material
389,1 nature, yet they do not happen because of this' (since to ask the question 'Why?' is indicative of cause, and since matter too is a cause, it would be in this respect that we can say it happened because of the matter); he quite reasonably adds 'except in the material sense',[408] for these[409] are causes, but not unconditional causes of the existence of the house, whereas the purpose is more truly a cause. But I cannot
5 understand why Alexander says that unconditional necessity is destructive of purpose. He says: 'Unconditional necessity is to be found where the end follows the antecedents of necessity.' Yet among eternal entities the end follows the antecedent of necessity (for example, the setting of the sun follows its rising, and the spring follows winter and is in turn followed by summer), and even so, each of these is for some purpose. For the winter comes first for the sake
10 of the spring, and the sunrise for the sake of the sunset.[410] Perhaps he understood necessity only in the case of the material and the result according to the matter, and so reasonably said that necessity destroys purpose. For if what happens does so only because of its matter, there is no final cause, and so there is no purpose. For even among natural entities what is always the case (if nothing prevents it) is
15 necessary, but this does not destroy purpose.

200a15 Necessity occurs in a parallel manner both in mathematics and in the products of nature to some extent. Because the straight is of this particular character, it is necessary that the internal angles of the triangle should add up to two right angles.[411] But the converse is not a necessary truth – although if it were not true, we would not have the straight. The reverse is true in the case of what happens for the sake of something. If the end is going to be, or actually is, of a particular character, then the antecedents will be, or are, of a particular character. Otherwise, just as in the case of mathematics there is no starting point if there is no conclusion, so here there can be no end or purpose, since that is the starting point – not of the activity but of the planning (in mathematics it is the starting of the reasoning, since there is no physical activity).[412] Therefore, if there is to be a house, these materials must be made, or be provided, or just be there.

In general there must be matter which is for some purpose (for example bricks and masonry for a house). Nevertheless, the end will not exist because of these materials except *qua* matter, nor will it come into being because of them. In sum, neither the house nor the saw will exist if their materials do not exist (masonry for the house, iron for the saw); nor will there be any mathematical starting points if the internal angles of a triangle do not add up to two right angles.

He reminds us of the character of conditional necessity by reference also to the reasoning of mathematics.[413] For there is conditional necessity in mathematics too, although it is not of the same character as that to be found in the products of nature. That is why he adds the phrases 'in a parallel manner' and 'to some extent'. For in the case of mathematics, if the conditions are confirmed, then too is the conclusion drawn from them. His example is: 'Because this thing is straight, in other words, since this sort of figure, viz. a triangle, is straight-sided, it is necessary that the internal angles of the triangle should be equal to two right angles.' And if this does not have <internal angles equal to two right angles>, then it is not a straight-sided figure. But if this has its three <internal angles> equal to two right angles, it is not *necessary* that it should be a straight-sided figure. For it is possible for the three <internal> angles of a four-sided figure, not just those of a triangle, to be equal to two right angles. This is more evident in the case of numbers. If 5 is added to 5 it makes 10; but if there is a 10, it is not always 5 + 5; for 9 + 1 = 10, 8 + 2 = 10, 7 + 3 = 10, 6 + 4 = 10; but without 10, there is no 5 + 5. Generally speaking in reasoning of this sort the affirmation of the consequence follows from the affirmation of the leading principle, but the affirmation of the leading principle does not follow from the affirmation of the consequence, but the denial of the leading principle follows from the denial of the consequence. For although the principle is not proved by the consequences, it is denied when they are denied.

In the products of nature if there is going to be, or if there already is, a house, it is necessary that the materials should be of a particular character; if they are not of this character, then there can be no house. But it is not the case that if those particular materials exist, then there must be a house.

It seems according to this reasoning that there is conditional necessity to be found both in mathematics and the products of nature.

The difference lies in the fact that in the case of mathematics the assumption is that if the prior is confirmed, then so is what follows. For the propositions or hypotheses are prior to the conclusion, and if the consequence is denied, so too is the prior. The converse is true in the material world. For if the end is necessary, then so too are the means to it; but even if the means to it exist, the end is not necessary
10 *tout court*. If there is a house, there must be masonry; but even if there is some masonry, there is no necessity for there to be a house. The reasoning would, however, be similar if one considered the end as a beginning in nature in that the matter exists for its sake. This is what Aristotle himself will again confirm from the field of art. For right at the outset one takes the needs and usefulness of the house into consideration and makes preliminary drawings, so bringing the materials into play. The conception of the end in view is the starting
15 point of the planning but not of the work; for there are two starting points, one of the planning, the other of the work, and it is the conception of the end which is the starting point of the planning. For since there is need of protection, a roof must be built, which requires walls, foundations and trenches. The end of the planning becomes the starting point of the work. But in the field of mathematics, where there is no physical work to be done there is a single starting point
20 in the planning; it is not an end or a purpose, but what points the way to the conclusion.

But Aristotle draws a closer parallel between mathematics and natural entities. For just as in mathematics, when the end is established, i.e. the conclusion which is based on the propositions which are laid down of necessity, the propositions which are the starting
25 points in them are not there of necessity because the same point can be proved by other means,[414] similarly, not even in the case of natural entities do the theoretical starting points follow from any necessity in the materials. For the end does not follow from the matter. In both fields, therefore, the theoretical starting point does not follow from a necessary precondition (in mathematics the theoretical starting points are the propositions, in the case of natural entities the end in
30 view); when these are established the next stage follows – in mathematics the conclusions, in physical entities the matter.

Since the example of the triangle was unclearly explained, because Aristotle merely pointed at the argument, we should note that it has been shown that the external angles of any straight-sided figure add
35 up to four right angles and that the internal and external angles of a triangle in total add up to six right angles.[415] Therefore since the triangle is the straight-sided figure to which Aristotle referred when he said 'this thing is straight-sided', and since its internal and external angles add up to six right angles of which the external
391,1 account for four, the remaining internal angles are equal to two. If

this is not the case, the external angles will not add up to four, since the total is six, and so this is not a straight-sided figure. But if the three internal <angles> are equal to two right angles, it is not always a triangle.

> **200a30** It is clear that necessity in natural entities is what is said to be the matter and the changes that take place in it. The natural scientist should state both causes, stressing the purpose, since this is the cause of the matter, but the matter is not the cause of the end – and the end is the purpose, and the starting point is in the definition and the account, just as is the case with the products of art: since the house is a particular kind of thing, then the materials must be what they are and must of necessity be available; and since health is something particular, then its materials must be what they are and must of necessity be available; similarly if a man is to be what he is, then his matter must be of a particular kind etc., etc.

He has shown that necessity in natural things resides both in the matter – as the early natural scientists thought – and also in the changes and alterations of matter, but not unconditionally in such a way that the end and in general the form follow from the matter of necessity, but conditionally, as has been said. He finishes off by concluding that 'the natural scientist should state both causes,[416] stressing the purpose' since this is more strictly <the cause> in that this is the cause of the matter. When something is established and the rest follows of necessity, then the former is the cause of the latter. So if, when the matter is established, the end does not follow from it of necessity, but if, when the end is established, the matter follows from it of necessity, then the end would be the cause of the matter. The end is the cause of the form too in cases where the form is different from the end. If, then, the end is the cause of the matter's existence, the natural scientist who is discussing causes will rightly talk about the end rather than about the matter.

Having stated that the matter is not the cause of the end, showing what he means by 'end', he goes on to mention the purpose which precedes it; for this is the cause and the starting point, but not the ultimate goal. But since in most cases the form is the purpose, he points this out and adds that this is what the purpose is, and that such an end, which is the form, is also 'the starting point taken from the definition and the account'. For this is the cause of the matter as it is in the case of the products of art. For having defined the form of the house, which is a shelter to protect against wind, rain and heat, we prepare our materials in the light of this definition. This is true also of natural entities. For if there is to be a man, the male seed and

the menstrual fluid in the woman's womb must come together; and if any other animal or plant is to be produced, the appropriate matter must be provided. This is what he means by the 'etc., etc.' Here too Aristotle is following scientific method and naturally concurs with his
30 teacher; for Plato says in the *Timaeus*,[417] when drawing the distinction between these two causes, the material and the one which perfects according to intellective prevision: 'What we have said so far, with little exception, has been an explanation of the workings of Intellect; we must now hold up for comparison what happens through necessity. For the creation of this universe was a mixed birth of
35 necessity and Intellect. But Intellect controlled necessity by persuad-
392,1 ing it to guide most of what was created to a better state. In this way and according to the same principles necessity was overpowered by sane persuasion, and this universe came together in the beginning. Therefore if anyone is to say truly how it was born in this way, he
5 must include the form of the wandering cause too.'

200b4 Perhaps necessity is in the account too. For when we have defined the task of sawing as being cutting of a particular kind, this cannot happen unless the saw has teeth of a particular kind. These teeth can only exist if the saw is made of iron. For even in the account there are some parts which are the matter of the account.

He has said that the definition and the account are the starting point and the cause of the matter, which represented necessity even among causes, since even matter is included in many definitions, sometimes
10 in potentiality in the definition of the form, as he himself suggests, and sometimes in actuality as in the compound.[418] It is, then, clear in the former case that the definition will no longer be the starting point and cause of the matter by including it in itself, which is why he adds 'Perhaps necessity is in the account too'. For in the latter case the end and the form are the starting point and cause of the matter. Such a
15 definition is indicative of the compound, so that it is stated that the natural scientist will give an account of both types of cause, but particularly about the end. In the case of natural entities this is the form and the definition, so that most of his discussion will be about the form and the definition. Yet on occasions he will talk about both types of cause in his definition, in cases where the matter is included
20 in the definition and the account. As Aristotle himself says elsewhere,[419] complete definitions of natural entities are given in terms of the form but include the matter, since the natural form includes matter and substrate. That is why on some occasions the definitions are taken from the matter, for example when we say that anger is a
25 boiling of the blood around the heart; on other occasions they are

taken from the form, as when we say that anger is a yearning for revenge, and on yet other occasions they are taken from the compound, as when we say that anger is a boiling of the blood around the heart due to a yearning for revenge.[420] We ought to note that it was from the definition and the statement of the essence that the matter of necessity made its entry. For in the field of art matter depends on the thoughts of the craftsman, which consist in their concepts of the 30 intended form according to the definition of the form. Therefore the definitions themselves single out the matter, judging its appropriateness from the formal concept. For just as he first takes the definition of the house, that it is a shelter for the use of men which requires timber and masonry for its erection, so he comes then to the mate- 393,1 rial.[421] Consequently, matter depends on the form and is posterior to it *qua* matter. If one were to define a saw and say: 'A saw is a tool for cutting timber or stone in a particular way, being fitted with teeth of a particular kind', it will be apparent at the same time that the teeth need to be made of iron. The doctor proposing an end, namely health, 5 which is the balance of hot and cold, dry and moist, will single out his material to suit this. This is the case with natural entities too. He seems by this method to show that the form or the end is the cause of the matter, since it is by the definition of the form that the conception of the matter is revealed. Even if it is clearly put, the matter will be in the definition, and then it will be a part of it as the 10 matter of the definition, but not the true cause.

This concludes my commentary on the second book of Aristotle's *Physics*.

Notes

Abbreviations

DK	H. Diels, *Die Fragmente der Vorsokratiker*, 6th edition rev. by W. Kranz, Berlin 1952
KRS	G.S. Kirk, J.E. Raven & M. Schofield (eds), *The Presocratic Philosophers*, 2nd edition, Cambridge 1983
LSJ	H.G. Liddell, R. Scott & H.S. Jones, *A Greek-English Lexicon*, 9th edition, Oxford 1940

Works of Aristotle

An. Post.	*Analytica Posteriora* (*Posterior Analytics*)
An. Pr.	*Analytica Priora* (*Prior Analytics*)
Cael.	*de Caelo* (*On the Heavens*)
Cat.	*Categoriae* (*Categories*)
DA	*de Anima* (*On the Soul*)
EE	*Ethica Eudemia* (*Eudemian Ethics*)
EN	*Ethica Nicomachea* (*Nicomachean Ethics*)
GC	*de Generatione et Corruptione* (*On Coming-to-be and Passing Away*)
Metaph.	*Metaphysica* (*Metaphysics*)
PA	*de Partibus Animalium* (*Parts of Animals*)
Phys.	*Physica* (*Physics*)
Top.	*Topica* (*Topics*)

1. *Phys.* 184a10-16.

2. Aristotle discusses the terms 'principle' (*arkhê*), 'cause' (*aition*) and 'element' (*stoikheion*) at *Metaph.* 5.1-3 and 12.4. The term 'element' often refers to the four corporeal elements (earth, air, fire and water), but here the 'elemental principles' are those of compound bodies, viz. matter and form.

3. cf. Philoponus *in Phys.* 194,4-15 for a similar résumé of book 1.

4. Book 2 of *Phys.*

5. It is typical of Aristotle's method to discuss and comment on the views of previous thinkers before proceeding to expound his own. Those who 'said that nature was matter' are not specified, but include such early natural scientists as the Milesian hylozoists and the atomists, while those who 'claim that it is form' include *in primis* Plato.

6. *Phys.* 192b1. Aristotle excludes from the argument the transcendent forms of Plato, and confines his discussion to immanent or enmattered forms, which he calls 'natural and perishable forms'. See Irwin (1988) 243-51.

7. cf. Aristotle *An. Post.* 89b24, where in our enquiries into the nature of anything we are urged to ask four questions: what sort of thing it is; why it is what it is; whether it exists; what it is.

8. The Greek terms for 'luck' and 'chance' are *tukhê* and *to automaton* respectively. I have kept to 'luck' and 'chance' throughout the translation. The distinction between them is explained by Aristotle in *Phys.* 2.6 below. Others translate *to automaton* as 'spontaneity', 'luck' or even 'the automatic'.

9. Aristotle is picking up the promise made at the start of book 1, where he says (184a13ff.) 'We think that we understand each thing when we understand its first causes and its first principles right down to its elements', and at the start of chapter 7 of book 1 (189b31ff.) 'The natural way of proceeding is to start from what is common, and then to move on to what is particular in each case'. He has devoted most of book 1 to a discussion of previous theories of nature, and now proceeds to expound his own views. It should be noted, however, that (according to Ross (1936) 499) book 1 was originally a separate essay, attached to the main work at a later date.

10. Although Aristotle's work is generally known in English as the *Physics*, the Greek term *ta phusika* has a wider connotation, and the English 'natural' corresponds more closely to it.

11. The scope of the enquiry is to be limited to natural objects only in so far as they are natural, i.e. no other aspects of them are to be investigated. The distinction between 'natural and non-natural entities' made in the next sentence is not only a distinction between different kinds of entities, some of which are natural and others non-natural, but also between different aspects of the same entities. An example of a non-natural entity is given as a just or unjust action – which would lie entirely outside the scope of the *Physics*. An example of an entity of which certain aspects are natural while other aspects are non-natural would be a bed; at 263,29-30 Simplicius quotes Eudemus as claiming that a bed is moved not in so much as it is a bed but in so much as it is made of a natural substance, wood. Thus the enquiry is limited in the first instance to those things which are *per se* natural, and in the second instance to those things which can be said to be natural only *per accidens*, in the way that a bed is natural only in the sense that it is made of a natural material, as Aristotle himself recognises at 192b19-20, where he says 'a bed ... has no innate tendency to change, but because it is made up *per accidens* of earth or stone or some mixture of them it has such a tendency derived from them in so far as it is made up of them'. See further Charlton (1970) 88-93.

12. Although the verb *huphistasthai* and its cognate noun *hupostasis* did not become technical terms until the time of Porphyry (3rd century A.D.), with the Platonist connotations of the Real Being of the Intelligible World, Aristotle uses the verb as synonymous with *einai* (to be). *Substantia* (substance, substantial being) is the direct Latin equivalent of *hupostasis*.

13. The broad distinction between natural entities and those which fall outside that description is clarified by reference to cause. We should remember that we have already been introduced to the form-matter analysis in book 1, and it might seem a possibility to apply that analysis here. But discussion on those lines is delayed until a more complex survey of cause has been conducted in chapter 3. Instead Simplicius offers examples of causes other than nature. Although he does not state explicitly that this is an exhaustive list, it is necessary to the argument that it should be, since he is 'discovering precisely what nature is' by a process of elimination. It is not until his remarks in chapter 8 (198b2ff.) that he discusses the causative role of nature.

14. There is an implicit distinction between things that are and things that come into being – a distinction rooted in Platonism but employed by Aristotle. It is the latter that are the proper concern of the *phusikos*, the natural philosopher.

15. For the distinction between practical and productive reason cf. *EN* 6.4,

1140a1ff., where Aristotle equates art with productive reason (1140a7ff.) and differentiates its products from those of nature (1140a15). Here Simplicius gives as examples of products of practical reason just and unjust actions, and as examples of products of productive reason a bed, a house and flute-playing – none of which *per se* falls within the scope of natural philosophy.

16. Reason linked with desire is the definition of choice in *EN* 3.3, 1113a10-11; 6.2, 1139b4-5.

17. Reason not so linked operates in the field of art, for example, the production of artefacts or the acquisition of a skill. This seems to be an inference of Simplicius' own, since Aristotle would disagree. The production of artefacts requires desire (*orexis*). At most Aristotle allows that it requires no deliberation: *Phys*. 2.8, 199b26, commented on at 385,10ff. below. I have chosen to translate the Greek word *tekhnê* in the traditional manner as 'art', although the scope of the English word is narrower than that of the Greek. 'Human design' might be a more acceptable rendering.

18. Chapters 4-6 are devoted to a discussion of luck and chance. Luck is the unexpected result of purposive action on the part of a human, judged by its outcome as either good or bad, as in the example given here by Simplicius, a windfall of money. Such a windfall must result from a purposive action, even though the purpose was something entirely different. Chance is a more general term which covers any event that is not the result of purposive action; at 197b27ff. Aristotle gives as his examples a portent in the form of a solar eclipse and the fall of a stone. The point is clear: neither a portent nor the way a stone falls of its own accord – even if it provides a good base for something – can be said to be lucky, although they are certainly products of chance.

19. cf. 197b17. Simplicius seems to be conflating two of Aristotle's examples – a tripod falling so as to form a seat, and a stone falling so as to hit someone.

20. Simplicius differs slightly from Aristotle in his presentation of natural entities; he lists them in decreasing order of complexity – animals, plants, parts of animals and plants, the four primary or simple bodies or elements (earth, air, fire and water) – whereas Aristotle does not apparently allow for parts of plants, and puts parts of animals before plants in his order.

21. Simplicius develops the distinctions drawn above in 261,12-17, first, that between nature and choice. Choice, which at 261,13 was said to produce vice and virtue, is not limited in its potential to produce either a good or a bad act, and it can produce an infinite variety of good and bad acts; nature is however 'programmed' to produce what is definite – an acorn can become nothing other than an oak tree. Cf. *EN* 1111b20ff. where choice is contrasted with wish (*boulêsis*); choice is limited to what is possible, while we can wish for the impossible.

22. Having an internal cause of change is Aristotle's criterion for distinguishing natural things from artefacts.

23. Having isolated the products of nature from the products of other causes, he now asks what it is that the former have in common. Again the process is one of elimination; perception is ruled out as belonging only to animals, as are nourishment, growth and reproduction (as belonging only to animals and plants, and possibly their parts; Simplicius expands on this point below with a quotation from Alexander). What is left is (a) the ability to change locally, provided that there is no impulse of the soul. (What he appears to mean by this last qualification is that the only sort of local movement common to animals, plants and the primary bodies is that which is not caused by an impulse of the soul; although animals clearly do move on some occasions through impulses of the soul, there are occasions when the local movement of an animal, such as falling or rising to the surface of the water, is not the result of such an impulse and is thus akin to the movements of the primary bodies) and (b) the potential for qualitative alteration, coming into and passing out of being, increase and decrease – changes which are common to them all. For Aristotle's discussion of such elemental changes see *Phys*. 1.7.

24. The quotation is attributed by Simplicius to the Peripatetic Alexander of Aphrodisias (3rd century A.D.). It is presumably from his commentary on Aristotle's *Physics*, lost except for newly discovered excerpts on Books 4 to 8 being edited by Marwan Rashed: see Rashed (1995) and (1996) and Harlfinger and Rashed (1997). The first part of the argument is rather strangely stated, but the point seems to be that since some of the parts even of the products of art are natural things, such as the stone and timber of a building, all the parts of something natural must themselves be natural. The second part of the argument (from 261,34) is reductionist: at the simplest level the four primary bodies are the basic constituents of all things, whatever their proximate causes. There is some little confusion in that Alexander does not specify what he means by 'parts' or 'natural bodies' (*ta phusika somata* in 261,34-5). For 'parts' Ross (1936) 500 suggests organs and tissues, and Aristotle later on gives a clue in his mention of flesh and bone, which would parallel the stone and timber mentioned here by Alexander. 'Natural bodies' would however more readily refer to the four primary bodies. The terms *hupokeitai* at 261,34 and *ta hupokeimena* at 261,36 could refer to either. But the principle remains the same – the products of nature and art alike are all ultimately reducible to natural bodies.

25. Simplicius now draws a further distinction between parts (*merê*) and components (*stoikheia* – a term often translated as 'element'). The parts of a statue are its head, feet etc., while its components are its stone or bronze. The former are unquestionably non-natural. But he goes on to say paradoxically that the components – whether or not we call them parts – of both natural and non-natural things are not natural; the reason he gives is that their matter (*hulê*) is non-natural, since it lacks in itself a principle of movement, the very guarantee (according to Aristotle at 192b14) of their being natural. The argument rests on Simplicius' belief that Aristotle posited, at the final level of analysis of natural things, a 'bare' matter, lacking all character and definition which, though mobile, lacks an intrinsic cause or principle of movement. Simplicius further interprets Aristotelian matter as a kind of extension: see Sorabji (1988) ch. 1. Such matter is, however, outside the scope of natural philosophy both because it lies beyond physical change and because it is entirely unknowable. Simplicius thus denies it a place in the argument here, and suggests that the natural philosopher should look to the proximate components and not go below the level of the four primary bodies. He seems to be suggesting that the natural philosopher should analyse (a) natural things such as animals and plants in terms of their parts (bone, flesh etc.) and their proximate components (the four primary bodies), and (b) the products of art in terms of their parts (the arms and head of a statue, the different areas of a house) and their proximate components (stone, timber) – therein lies the difference (262,12); the similarity lies in the fact that in both cases the proximate components are natural.

26. A distinction is being made between parts (*merê*) and components or elements (*stoikheia*). Strictly speaking the parts of a statue are the hands, feet etc., which are products of art, while its components or elements are the bronze, ivory etc., which are products of nature. A similar distinction is drawn between the parts and the components of a house. Simplicius is using the term *stoikheion* in the broader sense here, whereas at 262,11 it clearly refers to the four corporeal elements.

27. Simplicius' puzzle is discussed again at 286,20-287,25 and fuller notes will be given there. Aristotle might have been prepared to say that souls were one kind of nature, but after him the Stoics contrasted soul with nature, rejecting Aristotle's ascription of soul to plants and substituting nature as something different. Hence Simplicius and his contemporary Philoponus, in passages given in the notes below to pp. 286-7, can speak as if they take soul and nature to be distinct causes, although both working to the same end and in living things. See Introduction, 2-3.

28. Simplicius seems to be making a fine distinction here. The matter *qua* bronze of a statue is natural, whereas matter *qua* matter, in that it lacks the principle of movement *per se*, is not. He resolves the problem by taking the analysis of the products of both nature and art no further than the proximate matter – bronze, timber, stone in the case of the products of human design, the four elements in the case of the products of nature – ruling out any discussion of the ultimate substrate, bare matter. The reference is to Aristotle *DA* 434a22ff.

29. The reference is to 193a12 and 193b8. The problem raised here by Simplicius is that of the entities listed by Aristotle at 192b8 – animals, plants and the four elements – the two former are commonly said to be what they are not so much because of nature as because of soul. Aristotle's fuller discussion of the soul in *DA* states broadly that what is particular to humans is the rational soul, to animals the locomotive (here described as desiderative) and to plants the vegetative. (This is not presented as a tight schema: at *DA* 414a13 he mentions five faculties of the soul, nutritive, desiderative, sensitive, locomotive and intellective, and at 416a19 he add the generative, at 432b1 the imaginative, and at *EN* 1102b28 he talks of the vegetative and impulsive faculties.) What is important is that each part or faculty of the soul subsumes all those below it, and in that nature stands to inanimate bodies as soul stands to animate beings, all animate beings, in that they are said to exist because of soul, can be said to exist also by nature. Plotinus considers nature as effectively a fourth hypostasis below the One, Mind and Soul. For his discussion of this point see *Ennead* 4.4.13.

30. *DA* 412a27.

31. The verb supervene (*epigignesthai*) points the distinction between Aristotle, for whom soul is an immanent entity, and Plato, for whom it is transcendent.

32. Simplicius perhaps has Heraclitus in mind. See Kahn (1979) 238-43, 248-54.

33. Wehrli (1955) fr. 50. Eudemus was a friend and pupil of Aristotle, and was apparently a contender to become head of the school after Aristotle's death. In the event Theophrastus became the head, and provided the buildings known as the Peripatos, from which the school took its name.

34. The Greek word *kinêsis* can mean either change or movement. I have generally chosen the former. Other words in current use were *metabolê* (alteration) and *alloiôsis* (qualitative change). The Greek word *stasis* can imply either a cessation of change already taking place, or the total absence of change, rest. The former is the more appropriate translation in this context. Cf. Philoponus' discussion on this point at *in Phys. 2* 196,17-20 and 198,9-199,23 in Lacey (1993).

35. An example of a rock changing 'from within itself' is the growth of crystals; an example of a rock decreasing 'from within itself' is the degeneration of, say, marble.

36. This sentence has teleological overtones; a process of change is seen as moving towards a particular end dictated by the form in question. For example, an acorn will not go on growing *ad infinitum*, but only until it becomes an oak tree of a size determined by the form. But, as Alexander observes, some changes do go on *ad infinitum*, for example the rotation of the heavenly bodies – although Simplicius finds a way in which even these can be said to be at rest.

37. The Greek word for 'rest', in cases where there is no preceding motion, is *êremia*. The point Simplicius seems to be making is that the rotational motion of the sphere extends throughout its bulk as far as the axis, which is itself at rest. Since the axis is only notional, both terms, *stasis* and *êremia*, can be applied to it. But his final statement seems to be back-to-front, and it would be more appropriate for him to say: 'But not all rest is cessation – only that after motion is.' The Greek word order, however, is not in favour of such a translation; nor does Simplicius appear to use the terms consistently in what follows.

38. See Appendix: The Commentators for Porphyry. Porphyry's point is a refinement of Alexander's position in that a heavenly body is a *simplex* and is not subject to the same rules of change as a *complex* such as fire, which is, however, the most divine of the elements. Cf. Plotinus *Ennead* 3.6.6.41 and Aristotle *GC* 335a8-10.

39. For the dispute between Philoponus and others on the motion of fire see C. Wildberg, 'Prolegomena to the Study of Philoponus' *contra Aristotelem*' in Sorabji (1987) 205-8.

40. Presumably Simplicius is paraphrasing part of Alexander's Commentary on Aristotle's *Physics*. The emphasis seems to be on the phrase 'within themselves', and the point is that the stated sorts of change take place in all three types of entity listed because they are natural bodies, i.e. they have the source of change within themselves. Under other descriptions they will possess other faculties, e.g. plants *qua* plants possess the vegetative faculty.

41. There is a variance in the manuscripts known to Simplicius. The reading 'source' (*arkhê*) is found in the paraphrase of Aristotle's *Physics* by Themistius (4th century A.D.). Philoponus reads *hormê* (impulse).

42. Simplicius uses the term *holotês* (wholeness, completeness) several times in this work (e.g. 336,34; 347,10). It is the nature of a stone to fall downwards to find its proper resting place on the earth, the element to which it belongs, thereby 'gaining its unity'. Cf. *Metaph*. 1023b36.

43. See note 36.

44. Simplicius does not add the corollary 'and they can change without limit'; the sculptor can continue to add or take away clay *ad infinitum*.

45. For Aristotle's exposition of syllogisms see *An. Pr*. 1.1-7. Lloyd (1968) 118-19 summarises: 'Syllogisms of the first figure may be described as those in which something is predicated of the middle term in the major premiss, and the middle term is predicated of something else in the minor premiss .... Syllogisms of the third figure are those in which something is predicated of the middle term in both the major and the minor premiss.' Aristotle thought that science should be syllogistically expounded, but seldom does this himself. A later reference in Simplicius' text below shows that his follower Alexander had already begun to do it for him.

46. Lucian *Vitarum Auctio* 26.

47. Diels comments 'I do not know where', but Simplicius seems to be referring to *An. Pr*. 1.7, 29bff.

48. In so far as it is what it is, i.e. *per se*.

49. cf. *Phys*. 191b5ff.

50. Virtue is separable from the soul, and white from the surface, in that virtue and white can be conceived of as formally distinct from the soul and from the surface respectively; neither soul nor surface enters into the definition of virtue and white. By contrast, triangularity does enter essentially into the definition of isosceles.

51. Both primarily and *per se*.

52. The arguments in this paragraph are complex and compressed. The first stage is unproblematic: rational thought is part of the essential being of a man (Simplicius might better have said 'of a man's soul'), and is there without an intermediary; it is therefore part of him *per se* and primarily in just the way that triangularity belongs *per se* and primarily to the isosceles. Simplicius then introduces the idea of motion, and uses the distinction between a natural entity (the man) and the artefact (the ship) to show that whereas the former has its source of motion within itself *per se* and primarily, the latter does not. He begins by showing that the helmsman (the source of motion for the ship) is in the ship, but does not belong to it *per se*, since a ship without a helmsman is still a ship. Although he is in the ship, he is in fact an external mover; nor does he move it primarily, since he uses the ship's equipment, e.g. the rudder, as an intermediary. Simplicius makes the further point that in moving the ship the helmsman moves himself neither *per se* nor primarily in just the way that the rational part of the soul moves itself locally, when it moves the body, neither *per se* nor primarily – although when it moves itself according to its essence in its desires, impulses etc. it is moving itself *per se* and primarily. See Philoponus *in Phys. 2* 196,2 in Lacey (1993).

53. Simplicius is referring to the notoriously difficult chapter 5 of *DA* 3, 430a17ff., in which Aristotle suggests that mind can exist outside the body.

54. Alexander's interpretation is that the mind referred to by Aristotle is not the logical faculty in the human soul, but that of God.

55. cf. note 29.

56. This is the account of the soul that a Platonist would subscribe to. The point Simplicius is making is that there is a clear distinction to be made between soul and nature. Nature resides in the substrate body as an immanent entity, whereas soul is separable and transcendent. Simplicius is calling in Aristotle in support of his Platonist position, exploiting Aristotle's view that at least some intellectual activity is separable from the body.

57. *EN* 1140a10.

58. *Metaph.* 1032a25.

59. e.g. the bed *qua* bed falls to the ground when dropped *per accidens* – the argument of 192b8ff. above.

60. Syrianus was the teacher of Proclus, one of Simplicius' predecessors in the Athenian Academy.

61. i.e. from the movement natural to them, such as the upward movement of fire or the downward movement of earth.

62. To amplify his definition of 'to have a nature' Aristotle introduces the idea that any natural body, *qua* natural, is a compound substance which 'has' (a) a substrate in which its nature resides, and (b) its nature which resides in that substrate. He then goes on to show that however we define the compound substance – generally using two or more of matter, substrate, form and nature in the definition – it will always be the case that the compound 'has' the nature.

63. e.g. Antiphon, mentioned in the next passage.

64. Or possibly 'in the first book' (of the *Physics*) chapter 7.

65. *Cael.* 268a16.

66. Trans. W.K.C. Guthrie, *On the Heavens*, London and New York 1929.

67. *Cat.* 2a13.

68. cf. *Phys.* 1, 185a31.

69. *An. Post.* 89b21.

70. *An. Post.* 71a1.

71. The discussion is based on Plato *Timaeus* 67Eff.

72. 193b22.

73. cf. *Metaph.* 5, 1014b27.

74. Homoeomerous, i.e. homogeneous, substances are those that have no internal distinctions, each of whose parts, however much they are divided, remain identical in make-up to all the other parts.

75. See note 25.

76. Antiphon was a sophist of the 5th century B.C. who lived in Athens and wrote among other works one entitled 'Truth' (*alêtheia*), the one referred to here.

77. The Greek word for 'cause to germinate' (*phuein*) is cognate with the Greek word for nature (*phusis*), which is used also to denote germination.

78. Behind the argument here lies the nature-convention antithesis, formulated by sophists of the 5th century B.C. See further Guthrie (1971) and Kerferd (1981). Antiphon's point is that a bed is wooden by nature, but a bed only by convention, as is evident from its behaviour when buried.

79. A thing cannot survive losing its essence.

80. All the philosophers listed here are Presocratics of the 5th century B.C. For a full discussion of the early natural scientists see KRS.

81. The Greek word used here for 'form' is *rhuthmos* rather than the more common *eidos* or *morphê*. Cf. Aristotle *Metaph*. 985b16 and 1042b14.

82. Simplicius seems to be making a distinction between (a) natural entities, whose matter or substrate at every level, whether the proximate or the primary level, is its nature, and (b) products of art, which may have the same proximate and primary matter as natural entities, but which cannot be said to have that matter as its nature in that art has interfered in the chain. Thus the wood of a tree can in this sense be said to be its nature, while the wood of a bed cannot.

83. At Plato *Gorgias* 465B geometers are portrayed as presenting analogies in a form similar to Simplicius here: as A is to B, so C is to D. There could be an allusion to *Meno* 82Bff, where a slave boy (the Greek for 'slave boy' – *pais* – is the same as that for 'child') is questioned about a geometrical figure, and proportions are part of the discussion. cf. Proclus *in Tim*. 242B.

84. The distinction seems to be between a substance and its accidental properties (those which are said to be 'in' a substance at *Cat*. 1a20ff; the former persists through change, while the latter do not. See further *Phys*. 1.7.

85. Enmattered forms are the product of Platonism modified by Aristotelian thinking. For Plato the forms are separate entities, and it is only copies of them that inhere in the corporeal beings of the Sensible World. Cf. *Timaeus* 50C4: the copies of realities that pass in and out of the Receptacle. An Aristotelian form is an immanent rather than a transcendental entity, immanent in the objects of the world and separable only in thought. Neoplatonists bridged the gap by allowing for enmattered forms.

86. For Porphyry see Appendix: The Commentators.

87. Reading *phusin* not *phusis*.

88. 193b6ff.

89. Simplicius' etymology is suspect here. He derives the Aristotelian term *entelekheia* (perfect realisation) from the Greek words *hen* (one) and *telos* (end, completion). The first part of the word, however, is more likely to be nothing more than the intensifying prefix *en-*.

90. Simplicius seems to be using the term 'form' here both in the strict sense, form as opposed to matter, and in a broader sense to designate the compound of form and matter, the substance. See Charlton (1970) 70-3. Alternatively, we might translate: '... whether, when a man begets a man, it is the form that comes from the form, or rather the composite from the composite. Or maybe <the latter>. is his point. If this is so, then shape too is nature, since the shape also comes-to-be in the composite.' In which case there is no equivocation of the word *eidos*.

91. Throughout this passage Simplicius (like Aristotle) uses words cognate with *phusis* (nature), for example *ekphusis* (growth) and *phuesthai* (to grow).

92. The Greek verb *ginesthai* can be translated either as 'come-to-be' or 'be made'; the former here is the more appropriate in the case of the products of nature, and the latter in that of the products of art.

93. Natural change can be described in two ways, either (a) from matter to form, as in the case of the acorn growing into the oak tree, or (b) from opposite to opposite, as in the case of hot to cold.

94. The problem is discussed in *Phys*. 5 and in *GC* 1.3. See also *Phys*. 1.6-9.

95. *Phys*. 1.6-8

96. Since nature and form (in the sense of the substance which has reached completeness) have been identified, and since form is one of the pair of opposites 'privation and form', then a paradox is engendered: a form implies its opposite; but if the opposite is seen merely as a privation, then privation might appear to be a nature. Simplicius needs to show that privation is not just an absence, but absence of something, and thus 'a sort of form'.

97. *GC* 318b17ff. Aristotle suggests that of the pair 'hot and cold' the latter is a privation and the former a 'positive predication and a form' (*kategoria tis kai eidos*). Simplicius reminds us that we should not consider all process to be from worse to better – although his discussion here is concerned with those instances of change or growth which are.

98. *Phys*. 1.7. By 'the highest type of opposition' Simplicius seems to mean 'hot – cold', i.e. form – form, as opposed to 'hot – not-hot', i.e. form – lack.

99. cf. the descriptions of the warring of the opposites in Plato *Timaeus* 49Bff. and Aristotle *GC* 331a28ff.

100. *Cat*. 3b24. The problem is this: coming-to-be and passing away cannot happen absolutely – it was a maxim of Greek science that nothing could come-to-be or pass away absolutely. Aristotle's description of coming-to-be and passing away in *GC* and *Phys*. is of one thing changing into another either substantially, as water to air, or non-substantially, as uneducated man to educated man. In each case the analysis depends on the presence/ absence in a substrate of one or more qualities which are themselves further analysed in terms of opposites. Substantial change occurs when one essential quality is replaced by another. By describing pairs of opposite qualities differently as form and privation he seems to have engendered a problem here. He has shown that form is, in one sense, nature. If privation is a sort of form or a faint image of form (which is a different analysis from that into two opposites), then privation must have some share in nature – which seems a nonsense in this analysis. Further, since all coming-to-be and passing away is from opposite to opposite, and since there is no opposite to substance (which is complex, whereas a form is simple), it would appear that there can be no coming-to-be or passing away of a substance – which is what Simplicius says Aristotle claims is the only real coming-to-be and passing away. The explanation which follows tells us that there is no absolute coming-to-be or passing away of a substance, but only in respect of its formal element. He gives fire as an example. The problem is made more acute in that the discussion centres around plants, whose growth implies an improvement, an achievement of a goal. We have seen that nature in this sense is a progression to nature. It seems paradoxical, therefore, to suggest that the formal starting point of the growth, which is clearly inferior teleologically to the finishing point, and which can equally well be described as a privation, should be able to be termed its nature, even in the restricted sense suggested.

101. Simplicius is offering an alternative explanation, that the form is the substance. In this sense the form cannot have an opposite, and so its nature cannot have an opposite. Thus the possibility of privation being nature is ruled out. He concludes by suggesting coming-to-be and passing away for a form in this sense is according to the coming-to-be and passing away of the differentiae of the form. E.g.

fire as a substance-form is destroyed when its properties/ differentiae are changed. For forms as collections of properties see Spellman (1995) ch. 3.

102. There is a certain ambiguity in the way in which matter is designated. The proximate matter of a thing (e.g. the bronze of a statue) is on some occasions termed its primary matter, although this term is more normally applied to the bare matter, devoid of all qualities, which at least to a Neoplatonist is seen at the very end of the procession from the One, the bare matter of e.g. Plotinus *Ennead* 2.4.

103. The form which is inherent in the wood *qua* natural entity, which, when buried, puts forth shoots is the form of wood; under a different description the wood is merely the proximate matter of the bed.

104. Wehrli (1955) fr. 51.

105. 193a20ff.

106. *Phaedrus* 245E.

107. *Laws* 895C.

108. *DA* 413a20.

109. The Greek phrase *ex heautôn* seems to be ambiguous here. Things above nature (such as, ultimately, the Unmoved Mover) cannot be changed 'from their essential nature', while artefacts cannot be changed 'of their own accord'. Cf. *aph' heautôn* in 286,14 'by themselves'.

110. The Greek word for vegetative is *phutikê*, cognate with *phusis* (nature).

111. For the view that the nature of an animal or plant is identical with its soul for Aristotle, see Sorabji (1988) 222. But the relationship between nature and soul had been put in doubt by the Stoics, who, long before Descartes, rejected Aristotle's view that even plants have a soul, and ascribed to plants a *nature*, which they contrasted with soul. Hence Simplicius and his contemporary Philoponus can treat soul and nature as distinct causes, though both working towards the same end in living things. See Introduction.

112. The Stoics; see Fleet (1995) 126ff.

113. Diels comments *nescio ubi* ('I do not know where'). It perhaps refers to *Phys.* 8.4 255b13-256a3. Aristotle wanted the natural movements of the four elements up and down to be due in part to something external, a releaser, in order to establish the principle that whatever moves is moved *by* something and thus to make way for his unmoved mover, God. But Aristotle applies this principle even to ensouled animals, who are influenced by their environment (see Sorabji (1988) 219-26, 'Nature and God: Two explanations of motion in Aristotle'). It is an obstacle to Simplicius' attempt to distinguish nature from soul as a principle of being moved, if even animal souls are moved. But in another sense Aristotle argues that souls are unmoved: *DA* 1.3, 406a3; b7-8; 1.4, 408b5-8; 2.5, 417a31-b16; 418a1-3; *Phys.* 8.5, 258a7; a19. (Cf. 367,16-18 below.)

114. Diels gives *Phys.* 260a3, but the statement attributed here to Aristotle does not quite match that passage. Rather it applies more generally to his doctrine of the Unmoved Mover.

115. *Phys.* 192b20.

116. *Phys.* 255b29.

117. *Phys.* 255a24.

118. *Cael.* 284b33.

119. *Phys.* 199a19.

120. *Phys.* 199b32.

121. *Cael.* 271a33.

122. *Phys.* 199a8.

123. Simplicius seems to be slightly misquoting Aristotle here (or else he is reading from a variant text). The manuscripts have *proteron* (earlier), whereas Simplicius reads *heteron* (other).

124. The Neoplatonist 'creed'.

125. i.e. the cosmos: Aristotle *Cael.* 286b10.

126. Posidonius was a Platonising Stoic who was an influential figure in Rome in the 1st century B.C. Geminus was a younger contemporary. I have followed Diels' suggestion and added the definite article (*tês*) to go with *exêgêseôs*, 'his commentary' rather than 'a commentary'. See Edelstein & Kidd (1988) T42 (p. 33), T73 (p. 58) and F18 (p. 129ff.) for a fuller discussion of this passage.

127. I suggest a minor textual emendation, replacing *tropous* (methods) with *tropas* (orbits).

128. Wehrli (1953) fr. 110. Heraclides of Pontus was a Greek of the 4th century B.C. who studied under both Plato and Aristotle. He was briefly head of the Academy in 360/1 during Plato's absence in Sicily. See Gottschalk (1980) 58-69.

129. Platonists. I have printed Form with a capital 'F' since this phrase is effectively a name.

130. The criticism is levelled against those who fail to distinguish between qualities such as concavity, which can be abstracted from matter and are therefore separable from matter – as Aristotle allows, at any rate in thought (cf. *Metaph.* 6.1, 1025b33) – and those that cannot, such as snub[nosed]ness, which cannot be abstracted from their matter. The failure to make the distinction leads to major problems.

131. The craftsman referred to here can be taken either on a cosmic scale as the divine Demiurge in Plato's *Timaeus*, or on a more mundane level as e.g. the builder of a house – a favourite example of Aristotle's. That Platonic Forms exist in the mind of the Demiurge is a Middle Platonist view, reported already in Seneca *Ep.* 65,7.

132. 'Down here' and 'up there' are typical Neoplatonist terms used to refer to the Sensible and the Intelligible Worlds respectively.

133. For homonymity, i.e. ambiguity in words, see Plato *Phaedo* 78E, 102B, 102C, 103B, *Timaeus* 52A, Aristotle *Cat.* 1a, Plotinus *Ennead* 1.2.2.

134. *Metaph.* 1075a13.

135. Simplicius here quotes Aristotle as saying 'What then is it that is growing?' (*ti oun phuetai?*); the exact words at 193b17 are 'What then is it growing into?' (*eis ti oun phuetai?*).

136. Possibly a reference to *Metaph.* 7.11, 1037a29ff.

137. *Timaeus* 36E.

138. Simplicius is working hard here to reconcile Platonic and Aristotelian accounts of the forms, suggesting (and using Aristotle's own words) that Aristotle drew a distinction between mentally separable paradigmatic forms and their enmattered substantiations, giving them a causal role.

139. These people do not seem to be a clearly identified group. For insistence in Neoplatonism that cause need not be like its effect, but may instead be greater: see e.g. Plotinus 6.7.17 and Sorabji (1983) 315-16.

140. *Metaph.* 1072b20, 1075a11.

141. *Cael.* 292a18.

142. 193a31.

143. The Platonic Form, acting as a paradigmatic cause, belongs on a Neoplatonist list of six causes (see 316,23-6) which expands Aristotle's own list, in *Physics* 2.3, of four causes. Aristotle himself at 194b26 associates the formal cause with being a paradigm, but he is not thinking of Platonic Forms.

144. The duality of being both mover and moved.

145. i.e. the Platonic forms at the highest level in the Intelligible World.

146. e.g. snub[nosed]ness.

147. Matter is unknowable by direct experience and can only therefore be grasped 'by analogy'. This is true both for the Platonist (cf. Plato *Timaeus* 52B where the Receptacle, equated with matter by later Platonists, is said to be 'apprehensible without sensation by a sort of bastard reasoning', Plotinus *Ennead* 3.6.7 and Aristotle *Metaph.* 7.3) and for the Aristotelian (*Physics* 1.7, 191a8).

148. I have changed the punctuation here by closing the parenthesis in 300,5 not 300,4 (as Diels), and below in 300,11 by inserting a question mark.

149. Democritus was an atomist of the 5th century B.C.: see KRS ch. 15.

150. It is difficult to see of which verb Empedocles is the subject, of which verb Democritus is the subject, and of which verb both are. The quotation from Empedocles = DK 31B98 = KRS 373.

151. Anaxagoras of Clazomenae lived in the 5th century B.C.: see KRS ch. 12.

152. Plato reports, at *Phaedo* 97Bff., that Socrates felt a similar disappointment in Anaxagoras when searching for a non-mechanistic explanation of change.

153. DK 59B13.

154. i.e. not an internal cause such as nature.

155. DK 59B12. It is strange that the adjectives 'indefinite' and 'self-ruled' are in the neuter gender, suggesting 'something indefinite and self-ruled', while 'mind' and 'subsisting alone by itself' are masculine to agree grammatically with 'mind'.

156. 'Were all things together' is a quotation from Anaxagoras = DK 59B1 (*homou panta khrêmata ên*).

157. Simplicius is employing Neoplatonic vocabulary in this sentence.

158. Simplicius uses the word 'art' (*tekhnê*) here, whereas Aristotle uses 'science' (*epistêmê*). I have translated it as 'science' here to avoid confusion with the other appearances of *tekhnê* in this passage.

159. The argument is elliptical; in full it would be: it belongs to the science of building to know both the complete house (the end or goal) and the means to it (the roof and the walls, as parts contributing to the whole).

160. In Aristotle's analysis each process of change, both in nature and in the field of art, is taken to be a process complete in itself. He proceeds to exclude random change from the discussion.

161. Simplicius here uses the Greek word *morphê* (shape) rather than *eidos* (form) since the latter is conceptually more complex, although he uses the latter at 302,1.

162. The poet is identified by Philoponus, commenting on this same passage, as Euripides, although Ross (1936) *ad loc.* questions this attribution and suggests that it may be a line from a comic poet.

163. The Greek word is *skopos*, which has a range of meanings, of which 'aim' or 'target' is the most appropriate here. The distinction made here between objective and beneficiary is made in different terms by Aristotle also at *On the Soul* 2.4, 415b20-1.

164. These are generally considered to be different works, the latter not having survived. See Guthrie (1981) 83. Ross includes a range of fragments from *de Philosophia* in Ross (1955). Simplicius may here be referring to *EN* 1.1, 1094a3-6.

165. Simplicius is pressing the distinction between the two types of skill, but the upshot is that whatever the emphasis of the particular skill, its practitioner must know both the form and the matter.

166. This sentence is included in the previous *lemma* by Diels, but is not discussed by Simplicius until this point, I have therefore printed it here. For the statement that 'knowledge is in the category of the relative' see *Cat.* 7.

167. Simplicius seems to read this phrase as a question 'up to what point?' rather than taking the Greek word *tou* as indefinite, with the meaning 'up to a certain point'. See his discussion below.

168. As established in the discussion leading up to and including that of 194a36ff. The natural scientist will know matter in terms of the end imposed on it by the form.

169. See note 145.

170. Perhaps Simplicius is punning on the way Aristotle uses the word 'end' here.

171. Just as his enquiry about matter will be in terms of its end (see note 168), so too will his enquiry about natural or enmattered forms.

172. The Greek word *tis* (whose various forms include *tinos)* depending on its accentuation, can be either interrogative – 'who?', 'what?' – or indefinite – 'a certain'. If the accent is on the first syllable, then it is interrogative; if on the second it is indefinite (where the form allows a second syllable). Accentuation was unknown to Aristotle, and was introduced at a later date by Alexandrian scholars as an aid to the non-Greek speakers of the Hellenistic world.

173. See note 170.

174. See *GC* 2.10, 336a15ff. for Aristotle's view that the sun is the ultimate cause of coming-to-be for all natural entities in the sublunary world. Cf. also Plato *Republic* 6, 507Aff.

175. Alexander makes this inference presumably from the last sentence of the *lemma*.

176. Diels' punctuation is puzzling here, since by putting the accent on the second syllable of *tinos* he seems to be making Simplicius contradict what he has just said about the accentuation.

177. Alexander's interpretation is that Aristotle's statement is elliptical and in full would read: Just as the doctor and the bronze-smith know the matter of their trades only up to a certain point, so the natural scientist will enquire about form only up to a certain point.

178. The 'different reading' quoted here by Alexander adds the word *heneka* (for the sake of) after *tinos*, giving a different meaning: 'up to <knowing> what they are for'.

179. Themistius: see note 41.

180. See note 176.

181. Plato shares Aristotle's teleological concern in *Phaedo*. See David Sedley, 'Teleology and myth in the Phaedo' in Cleary and Shartin (1991) 359-83.

182. *Phaedo* 97C.

183. In the *Metaphysics* in general.

184. 184a12.

185. Cornford (1957) in a footnote to p. 128 (pointing to 195a29) explains this last phrase as 'the wider class to which these terms belong; e.g. bronze is a species of matter; statue, a species of image'. The Budé edition however has 'ainsi que les genres de l'airain et de l'argent' ('and likewise the genera of bronze and silver').

186. The Greek verb is *existasthai*, used by Plato and Plotinus of an entity which 'departs from its nature', i.e. ceases to be what it was, loses its essence.

187. The reference is to Thales of Miletus: see further KRS 76-99.

188. Unlike art, which can go on working on its matter *ad infinitum*.

189. 'Each of them' refers to (a) the first principle which is working towards the end in view, and (b) the matter at any stage of the process, already partly enformed. Between them they move the process on towards the final completion. Simplicius uses the example of grain, shoot, stalk and ear a little later on.

190. It is not clear just what sort of marionette show Alexander has in mind, but it would appear to be one in which one doll sets a second in motion, then the second sets a third in motion, and so on – a sort of 'domino effect'. Cf. Aristotle *MA* 701b2; *GA* 734b10; 741b9.

191. In that instantiations are like, but not identical to, the forms of which they are instantiations.

192. This is the third time in this passage that Alexander has used this expression, taken from Aristotle. It reminds us of the distinction made at 270,35 between what is by nature and what is according to nature. Only that which has reached the completion of the form, 'if nothing prevents it', can be said to be according to nature; things that fall short of such completion exist only by nature.

193. Material necessity: i.e. the matter necessitates certain by-products which do not themselves serve a purpose.

194. i.e. a Platonic form.

195. 'Essence' translates (as in the *lemma*) Aristotle's phrase *to ti ên einai*, lit. 'what it is to be'.

196. According to Alexander, the form of human which characterises the father-producer servers as a model for producing the same form in the child-product, just as the form of a house in the mind of the architect-producer serves as a model for producing the same form in a new house-product. Simplicius replies that in nature, which is not conscious like an artisan, the producer's form serves as a final cause, but not as a model.

197. The Greek word is *poiêtikos*, which can be rendered either as 'productive' or 'efficient'.

198. We should remember that Simplicius has ruled out any idea of Platonic forms in this passage, and is confining his discussion to enmattered or natural forms, which he sees as the end point in a process of change, and which seem to differ very little from the compound entity in its perfected state. The point is that there must be, according to Simplicius, a single cause producing a single perfected form; thus there is a difficulty in the case of a process where there are recognisable mid-points such as in the example given of grain, shoot etc., where each stage appears to stand as form/ actuality to its prior and as matter/ potentiality to its posterior.

199. Having complained that Aristotle's analysis, at least as Simplicius understands it, does not allow for continuity in the perfection of every natural form, Simplicius, in impassioned language, finds the answer in his own Platonic conception of cosmic order and planning – a noteworthy example of the cross-fertilisation of the two different schools. The reference back is to 287,14; cf. 314,14 below.

200. There was much debate in the ancient world about the roles of father and mother in reproduction. In the view of some the mother merely provided a place for the embryo to grow, and nourishment for it – see Cornford (1937) 187 for a list of sources for this view. However, the discovery of the female ovaries by Herophilus shortly after Aristotle's death gave support to the other view – see von Staden (1989) 230-4, Fleet (1995) 291 and Balme (1990) 20-31.

201. The Greek word that I have translated as 'creative rational principles' is *logos*, a word which has one of the longest entries in LSJ. Here it is reminiscent of the Stoic term *spermatikoi logoi*. See Urmson (1992) 120 n. 31; Sorabji (1983) 303. The problem here is that nature appears to be both rational and irrational. Simplicius' answer is that reason permeates the cosmos, and that nature, in that it does not operate in a random or disorderly manner, is at heart rational.

202. 194a29.

203. Alexander.

204. The Greek term for self-contemplation is *eis heauton epistrophê*, another familiar Neoplatonic term to express the turning inwards of a hypostasis to contemplate what is prior to it, e.g. soul turns inwards to contemplate mind, the second (Plotinian) hypostasis. See Lautner (1994).

205. *GC* 333b11.

206. This rather obscure statement seems to refer either to the activity of the demiurge in Plato's *Timaeus*, or in broader terms to the productive activity of intellect. Whatever the case, we are in a broadly Platonic framework where Platonic forms have been brought back into the reckoning. The term 'shining forth' (*ellampsis*) at 314,13 is a common Neoplatonic image.

207. Simplicius, by using the term 'definition' (*horismos*) here and at 314,34, implies that he understands Aristotle's use of *logos* (account) at 194b27 in this sense. Cf. *Metaph*. 7.5 1031a12.

208. Again the *lemma* was included in the previous Aristotle passage, but is discussed only here.

209. I have added '<or efficient>' here to indicate the more usual way of designating *to poiêtikon aition*; I have however kept to 'productive' throughout this passage in order to stress the etymological connection between the adjective and other words based on the same root, e.g. *poiein* (to produce).

210. See note 37.

211. 'Instrumental cause' is regularly counted by the Neoplatonists as one of the main types of cause. See 316,23-6.

212. Just as the shoot was the material cause of the stalk.

213. Plato did not, of course, list the causes as systematically as this statement implies. Simplicius is reading a lot into the passage at *Timaeus* 46Cff.

214. On joint causes see Introduction, 3-4.

215. The single genus is the question 'Why?', and the four causes relate to the four specific questions which comprise the genus.

216. *Politicus* 270A.

217. Familiar Platonic terminology for the contents of the Intelligible World; cf. *Timaeus* 52A. But it was Aristotle who insisted in *Physics* Book 8 against Plato that the primary mover must be unmoved, as is his divine unmoved mover, rather than self-moved, and Simplicius acknowledges here that the unmoved cause is prior to the self-moved.

218. *DA* 406a2ff., 408b15, 408b30, 411a25.

219. This is the case in both the Platonic and the Aristotelian account. For Aristotle the unmoved mover is God, who keeps the heavens (the everlasting intermediaries) in motion, so causing all movement and change in the universe. For a Platonist the unmoved mover is Intellect, who by self-contemplation, i.e. the contemplation of the contents of his own mind, the forms, is responsible for the contents of the Sensible World.

220. The imaging is the means by which form ('that which is participated in') is conveyed towards matter, and by which matter ('that which participates') gains its definition so as to become a natural generated form. Cf. Plato *Timaeus* 50C 'the images of realities passing in and out [of the Receptacle]'. The original is unaffected by the image.

221. Since nature in its details clearly works for a purpose, 'for the best', it would be absurd to suppose that the overall workings of nature were random. The thinking of both Plato and Aristotle is teleological, although both allow for a random factor that can interrupt or thwart the workings of nature, e.g. the Errant Cause in the *Timaeus*, and Aristotle's frequent disclaimer 'If nothing prevents it'.

For the (impossibility of) the primary cause itself being a product of chance see Plotinus *Ennead* 6.8.7.

222. Both Platonic Intellect and (in at least one passage) the Aristotelian God are eternal (*aiônios*), in the sense of existing outside time, whereas the universe which they produce and change is everlasting (*aïdios*), existing within time.

223. Simplicius seems to be misquoting Aristotle, who at 195a5 used the term 'many causes', and at 195a8 'some things'.

224. The Greek word for 'destruction' (*anatropê*) can also be translated as 'capsizing', appropriate here. Aristotle uses the same arguments and examples at *Metaph*. 1013a24ff.

225. 191a6.

226. Formal, final and efficient can often be subsumed under 'essence', as Simplicius points out below.

227. Simplicius is exploiting the double meaning of the word *stoikheion* ('letter' or 'element').

228. The 'Peripatos' is the name of Aristotle's school.

229. *Timaeus* 50B. The reference is to Plato's Receptacle, which is treated by Plato as space, but by Aristotle as a concept of matter.

230. Aristotle thought of menstrual fluid as the matter of the embryo which is en-formed by the male seed.

231. i.e. we should not take the Greek word *panta* (all) as the object of the verb 'acts', but in apposition to 'the seed', 'the doctor' etc.

232. *EN* 1094a2.

233. Wehrli (1955) fr. 52.

234. Diels has the definite article (*toû*), 'the good', but I prefer to read the (unaccented) indefinite *tou*, 'some good'.

235. Socrates chose death in order to avoid the greater evil of breaking his contract with the Laws and Constitution of Athens (*Crito* 50Aff.), just as many Stoics under the Roman Empire chose 'the rational departure from life' (*rationalis e vita excessus*). But Simplicius seems to be taking a harder line against suicide; cf. Plotinus *Ennead* 1.9. 17-19: 'But if we are to have in the other world the same standing that we leave this with, then we should not detach ourselves from it if there is any way forward'. The topic of suicide was commonplace in late Neoplatonist commentaries on Plato's *Phaedo*, and in the definitions of philosophy which preceded commentaries on Aristotle. Both discussed the definition of philosophy as separating the soul from the body or preparing for death. The *Definitions* of David have been translated from Armenian and the commentary of Olympiodorus on the *Phaedo* has been translated by L.G. Westerink, whose notes *ad loc.* are invaluable. 1.8 supplies five circumstances recognised by the Stoics as justifying suicide and, like Simplicius, Olympiodorus takes a more modified view. For further discussion, see J.M. Cooper, 'Greek philosophers on euthanasia and suicide' in Brody (1989) 9-38; Griffin (1976) and (1986); Sorabji (1993) 146-7.

236. 194b23.

237. 194b28.

238. The Greek word *diaphora* is often translated as 'distinction' or 'differentia'; but since Aristotle has so far made only a single distinction between nearer and more remote causes, I prefer here to translate the Greek phrase *tas duo tautas diaphoras* as 'these two species', i.e. the nearer and the more remote.

239. Milo, who lived in Croton in Southern Italy in the 6th century B.C., was an athlete famous for his physique and thus a popular model for sculptors.

240. Galen was a philosopher and doctor of the second century A.D., and a copious author. He was particularly influenced as a philosopher by Plato and as a doctor by Hippocrates. This reference is to a work entitled 'On the Doctrines of Hippocrates and Plato' 5.449 (Kühn). Simplicius is perhaps hinting that Aristotle's analysis of causes is as well articulated as a Polyclitan statue. For the Polyclitan Rule, see Pliny *Natural History* 34.55.

241. Simplicius' comment draws attention to Aristotle's explicative use of the word *kai*.

242. We should keep in mind the distinction between the builder who is (at certain times) only potentially building, designated here by the Greek word *oikodomos*, and the builder who is actually engaged in building, designated by the participle *oikodomôn*. See Ross (1936) *ad loc.*

243. The Stoic containing or cohesive cause (*sunektikon*) sustains things by holding them together and is also a sufficient cause of their behaviour. It may be because of the latter function that Simplicius uses the term, or he may be thinking in a non-Stoic way of a containing cause as incorporating other causes. See Introduction, 3.

244. 241b4ff.

245. i.e. as cause to effect.

246. In the previous sentence Simplicius has used two different versions of the phrase denoting potentiality, (a) *kata dunamin* (according to potentiality), and (b) *dunamei* (in potentiality). Alexander's comment seeks to explain the distinction between the two terms.

247. The question raised at *Cat.* 7b33 is whether perceptibles can still be said to exist if there is no perception of them. Philo the dialectician seems to have been the person who put the question in the form: 'Is the shell at the bottom of the sea visible?' Simplicius refers to this formulation again at *in de Anima* 17.29, where the point at issue is whether an act of the soul can be said to be separable if it is never in fact separated. See Sorabji (1983) 90-3 for a fuller discussion.

248. DK 31B53 and 59.

249. DK 68B167.

250. Wehrli (1955) fr. 53.

251. See *EN* 1140a19 for this quotation said to be from Agathon (Snell (1986) fr. 6 p. 163). Cf. *EE* 8.1 for the role of wisdom in good fortune.

252. This question had been ruled out in the case of nature at 193a3ff.

253. See note 45. The second figure is summarised by Lloyd (1968) 118: 'Syllogisms in the second figure are those in which the middle term is predicated of something else in both the major and the minor premiss.'

254. The Stoic Chrysippus devised five basic indemonstrable inference schemata, as described in Sextus Empiricus *Outlines of Pyrrhonism* 2.157f. and *Against the Professors* 8.224f., cf. Cicero *Topica* 54-7. See Kneale (1962) 162-3. The second of these is: 'If the first, then the second, but not the second, therefore not the first.'

255. Simplicius quotes a sentence from *Philebus* 19B which, after the Greek fashion, contains negatives which reinforce, rather than cancel, each other.

256. The death of Hector is described at Homer *Iliad* book 22. Adrastus was a Phrygian, who having unintentionally killed a man, fled to Croesus. While out hunting he unintentionally killed Croesus' son Atys, and consequently committed suicide. The story is recounted by Herodotus at *Histories* 1.43.

257. In Anaxagoras' view Mind was the constant causal factor in the cosmos; for Empedocles Love and Strife alternated, and for others one or more of the four elements predominated.

258. Wehrli (1955) fr. 54a; the story is of the death of Aeschylus – almost certainly a fiction.

259. The references to Empedocles are DK 31B53, 59, 98, 83, 75, 103 and 104.

260. 196a26 – this appears to be a paraphrase rather than a quotation.

261. 196a29.

262. Aristotle at 196a37 has *tout' auto* ('this very thing'), whereas Simplicius reads *tout' autôn*, where *autôn* (genitive plural = 'their') is best taken with *epistaseôs* ('close attention').

263. Wehrli (1955) fr. 55.

264. 'Fortune' might perhaps be a better translation of the Greek *tukhê* at this point, but for consistency I have kept to 'luck'.

265. *Laws* 709B.

266. e.g. Orphic Hymn 72 = fr. 204 Kern (1922).

267. There is a textual problem here (see Diels' *apparatus criticus*). I have omitted the phrase 'through art' (*dia tês tekhnês*) after 'choice' at 335,19. Diels suggests *idia tês tekhnês* ('particular to art'). The references cited by Diels are *EN* 1139a34 and *Topics* 116a11 (tentatively).

268. See note 45 and Ross (1936) 517.

269. Chance (*to automaton*) is distinguished from luck in *Phys*. 2.6, below, as covering coincidences which are not 'in accordance with choice'.

270. Wehrli (1955) fr. 56. Many of the words used by Eudemus are derived from the Greek word for 'luck' (*tukhê*), but this is lost in translation. I have rendered them variously as 'luck', 'fortune', 'achieve'.

271. See note 42.

272. The 'essential description' (*logos*) is that the man goes to the market in order to shop, or the stone falls because it is seeking its natural place. The two examples are taken from the fields of art and nature respectively.

273. As discussed in *Cat*. Only a substance (*ousia*) can exist *per se*; members of other categories, such as qualities, exist *per accidens* in that they can exist only in a substance; but they can be essential qualities of that substance, e.g. heat is an essential quality of fire, and is therefore present *per se*.

274. This point is explained by Sorabji (1980) 5-6, but it is there (ch. 1) given a further significance. There is no *per se* cause of lucky occurrences, for they are coincidences which lack a *per se* explanation and hence lack a *per se* cause. To take the present examples, the existence of a house is not a coincidence, and it has a *per se* cause. But finding treasure when planting is a coincidence and has no *per se* cause. Its *per accidens* cause, e.g. wanting to plant, is a *per se* cause only of digging or planting. On Sorabji's interpretation, Aristotle eventually decides, not in *Phys*. 2 but in *Metaph*. 6.3, that coincidences lack any genuine explanation or cause at all.

275. e.g. coming to shop when his debtor was there and in funds was luck; getting his debt back was the *outcome* of luck.

276. There are an indeterminate number of *per accidens* efficient causes, just as there are an indeterminate number of attributes of the *per se* efficient cause; the builder can be fair-skinned, artistic etc.

277. Being deformed, e.g. six-fingered, is not the outcome of luck, because we would not choose to be six-fingered. For 'ox-faced' see Sophocles *Trachiniae* 13. Cf. 'man-faced offspring of ox' in Empedocles (DK 31B61).

278. But they are the outcome of chance.

279. Simplicius is referring back to the discussion at 339,4ff.

280. Alexander, says Simplicius, is looking at the process of causation from the wrong end. Although a shopping trip to the market place could produce an indeterminate number of lucky outcomes, and it could be said that any one of these lucky outcomes is the *per accidens* final cause of the shopping trip, Aristotle's focus is on the analysis of the particular lucky outcome which consists in the recovery of the debt, which can have any number of *per accidens* efficient causes, all of which must have been intended for some other outcome.

281. Simplicius is saying that although luck can never be a *per se* cause when considered as part of a whole process, within a narrower conceptual framework it could be said that luck is the *per se* cause of a piece of luck *qua* piece of luck.

282. The Greek idiomatic expression for 'anything that chances' is *to tukhoun*, where there is a clear semantic link with *tukhê*.

283. Simplicius implies that the haircut was not part of the therapy, and so was only *per accidens* part of the process of recovery.

284. The Greek prefix is *eu-*, and there is no simple equivalent in English which can render this prefix in the words *eutukhia* (and its verbal equivalent *eutukhein*), *eukleês* and *eumathês*. I have used 'god-' as the best equivalent, although it should be noted that there is no religious connotation implied.

285. This is a difficult passage, open to various interpretations. Modern commentators – Ross (1936), Wicksteed and Cornford (1957) and Charlton (1970) – agree that Aristotle's argument at 195a28ff. (after a distinction is made between good and bad luck on the one hand, and a godsend and a disaster on the other) is as follows: to miss a disaster by a hairsbreadth is in fact to gain a godsend, since the mind thought that the disaster was already present, and a narrow miss *seems* to be no different from actually suffering the disaster; therefore the relief at finding out the truth is a godsend; and correspondingly to miss a godsend by a hairsbreadth is a disaster. Ross (1936) *ad loc.* summarises: 'To come within an ace of a great evil or a great good is counted good or ill fortune respectively, because we think of the great evil or good as having happened to us and then been lost.' Aristotle gives no examples, but we might consider an example of the former to be the condemned criminal, awaiting execution as something inevitable, who counts it a godsend when he is reprieved at the eleventh hour; correspondingly to miss the lottery jackpot by a single digit is considered a disaster. This interpretation assumes that the phrase in the Aristotle text *para mikron labein* means 'to miss by a hairsbreadth'. But the phrase is ambiguous, and can mean 'to achieve by a hairsbreadth'; this is how Simplicius appears to take it in 341,28-32 as 'a first interpretation' favoured by Porphyry and at least entertained by Alexander. I have translated accordingly. See LSJ *s.v. mikros* III.5(c). There are four points to note here. First, Simplicius does not use the phrase *para mikron labein* in this passage, but *para mikron ekhein*. Secondly, he prefaces his rendering of Aristotle's argument with the word *dunamei*, a well attested meaning of which is 'in effect', to go with the verb *sullogizein* which here seems to bear the meaning 'to syllogise'. Thirdly, in this interpretation Simplicius seems to have assumed that in the second sentence of the *lemma* Aristotle has been careless in his word order and has equated achieving a great evil by a hairsbreadth with a disaster, and achieving a great good by a hairsbreadth with a godsend. Fourthly, Simplicius may well at 344,1-3 be saying that *para mikron* in this case is to be understood 'not in the sense of coming close but being prevented, but in the sense of achieving the whole good or evil by a hairsbreadth' – but the wording is ambiguous even here. Simplicius, like Aristotle, offers no examples; we might imagine (if this is his meaning) an example such as the drowning man who reaches dry land with his last gasp and so survives. I am grateful to Professor David Sedley for his help with this note.

286. Then at 344,6ff he offers another interpretation which he says Alexander favours. Again the Greek is ambiguous, and could be taken in the way I have suggested in the translation, or: 'we call unlucky those who might have missed by a hairsbreadth some great evil, and lucky those who might have missed by a hairsbreadth some great good etc.'.

287. Understanding *an* in both clauses.

288. Diels suggests that the 'near gods' are Pollux and Castor, but says he has not heard them called this elsewhere. They were popular deities in Sparta, and the King always took an image of one or the other into battle (Herodotus *Histories* 5.75). They were widely believed to help those in danger, particularly at sea (cf. Propertius 2.26.9). See further Antonaccio (1995).

289. Retaining the *ouden* of the manuscripts, which Diels brackets.

290. The Greek word is *haplôs*, generally translated as 'simply, absolutely'.

291. By *endekhomenon* at 197a34 Aristotle probably meant contingent, but the translation is in terms of possibility, since Simplicius says the word is applicable even to what is necessary.

292. Simplicius has added the word 'only' to Aristotle's text.

293. The Greek phrase is *ta prakta*, lit. 'things to be done', but used in this more specific sense at *EN* 1094a19, 1095a16, 1097a22.

294. I have translated the Greek word *eutukhia* as 'good fortune' here, rather than 'godsend' as above.

295. *EN* 1153b21ff.

296. Protarchus was a pupil of Gorgias and appears as an interlocutor in Plato's *Philebus*.

297. Simplicius does not specify what he is referring to in the phrase 'that fact', but he must mean some genuine case of luck.

298. The Greek verb is *paskhein*, which together with the verbs *poiein* and *prattein* (and their derivative nouns and adjectives) present the familiar passive-active antithesis. The point is that the stones are purely passive and strictly speaking can suffer the outcome of chance, but not the outcome of luck, which is reserved for agents.

299. This appears to be a reference to the tyrant of Pherae, Jason, nicknamed 'the Thessalian Prometheus'. Cf. Cicero *de Natura Deorum* 3.28, Pliny *Natural Histories* 7.51 and Valerius Maximus 1.8

300. The story is told of the 4th century B.C. painter Apelles, the court painter of Alexander. It is cited by Sextus Empiricus in *Outlines of Pyrrhonism* 1.28.

301. The similarity lies in the fact that neither the agent (e.g. the painter or sculptor) nor the inanimate instrument (e.g. the chisel or sponge) join deliberately in producing the lucky outcome. 'Them' seems to refer to the stones discussed above.

302. At 197b34 Aristotle says: 'For when something occurs contrary to nature, then we say that it has occurred not so much as the result of luck, but rather of chance. But even this is somewhat different, since the cause of the one is external, of the other internal.' See note 316.

303. In that he would choose to regain his horse if he had the choice.

304. The distinction is between outcomes that are purely and simply what we would have chosen if we had had the choice (*haplôs haireta*) such as the retrieval of the horse, and those where some choice was made (*kata proäiresin*) such as going to the market to shop.

305. Bad luck too falls under the same analysis.

306. Either commonly as in everyday speech, or in the sense that we use a common expression for both luck and chance.

307. Aristotle, not alone of Greek philosophers, is guilty of suspect etymology. Here he suggests that Greek word *automaton* ('chance') is made up of *auto* (it) and *matên* ('to no purpose'). In his mind the connecting link is that chance follows on an event that is 'to no purpose', i.e. fails to achieve the outcome of which it was to be the *per se* cause. If the original event which fails to achieve its goal is directed by human intention, then the occurrence which follows is the outcome of luck. If there is no such intention, then the outcome is that of chance. Cf. note 89.

308. A stone falls to no purpose when the blow on the head, which would in other circumstances require a deliberate act of throwing by someone to make it purposeful and not the outcome of chance, does not happen because of any deliberate act, but because of a *per accidens* cause; the *per accidens* cause of the stone's hitting someone on the head is its natural downward movement (which is the *per se* cause of its finding its appropriate place), and the downward movement was to no purpose as regards the striking the blow.

309. i.e. coming to rest in its proper place.

310. Because the fall occurred to no purpose in respect of the blow struck, as Simplicius says at 350,31.

311. The stone's hitting someone supervenes on the fall of the stone, which took place to no purpose as far as what supervened (the blow on the head) is concerned; but since the blow might have been the result of deliberate aim, then the blow is the outcome of chance.

312. Rather than 'the outcome of chance'.

313. *Iliad* 2.408.

314. *Iliad* 5.749.

315. *Phaedo* 58A.

316. Aristotle's meaning is unclear. Charlton (1970) 110-11 sees the contrast as being between an outcome of chance, whose cause is external, and an occurrence such as a deformity, which is contrary to nature (although it is still 'by nature') and whose cause is internal.

317. Both these events – the retrieval of the horse and the way the stool landed – would move from the category of chance outcomes to that of lucky outcomes if it could be shown that they resulted from intended actions (although intended for a different purpose).

318. The Diels text has *elegen* ('he said'), but since these are Simplicius' own words above at 347,16 I suggest emending to *elegon* ('I said').

319. e.g. a physical deformity such as having six fingers.

320. He has just shown how going into the market-place may be the *per accidens* cause of finding one's debtor (i.e. if that was not the intention); he now shows how it can equally well be the *per se* cause (if that was the intention). But in the case of the horse the intention on the part of whoever caused the horse to find its way home was not that the horse should find its way home, but was some other intention; therefore this was a *per accidens* cause of a chance outcome for the horse and a lucky outcome for the owner.

321. The argument here, as in the Aristotle passage, is compressed and elliptical. It is: *per accidens* causes are posterior to *per se* causes; *per accidens* causes themselves have causes, which must ultimately be *per se* causes. The most ultimate causes are mind and nature; therefore, even if the universe has as its *per accidens* cause luck and chance, it must ultimately have a *per se* cause; since the universe is the first of all natural entities and is rational and eternal, its cause must be mind and nature.

322. *Cael*. 1.10-12, 279b4-283b32.

323. *Laws* 888Aff.

324. *Laws* 889B.

325. trans. A.E. Taylor, *Plato: The Laws*, London 1960.

326. *Laws* 896C.

327. The Greek word is *tukhê*.

328. Wehrli (1955) fr. 58.

329. A reference to a long-running debate among ancient philosophers. Aristotle believed that the account given in Plato's *Timaeus* was of creation in time – a view strongly contested by most Platonists – e.g. Plotinus *Ennead* 3.7

330. See note 307.

331. See above on 195b31ff.

332. See note 322.

333. *Iliad* 2.408.

334. The Greek verb *tukhein*, cognate with the noun *tukhê* (luck), can either bear the meaning 'to happen upon' or merely 'to happen'.

335. 'Hitting upon' is my rendering of the Greek verb *epitukhein*, again cognate with *tukhê*. Throughout this passage the argument depends on such words – happening upon, hitting, hitting a likeness etc. However, I feel that to force some form of the English word 'luck' on each occasion makes for a stilted translation.

336. Wehrli (1955) fr. 57. The Greek terms are *eutukhia* and *atukhia*, earlier translated as godsend/ good luck and disaster/ bad luck.

337. The language here is highly Platonic.

338. 196b6.

339. The term *sunaition* (joint cause) is used by Plato at *Timaeus* 46C and by Simplicius above at 316,25. It is recognised as a distinct kind of cause by the Stoics: *sunaitia* need each other in order to produce the effect. On Stoic distinctions among causes, see Frede (1980).

340. The Stoics distinguished the cooperating (*sunergon*) cause as intensifying the effect or making it easier to achieve.

341. All the words marked † are in Greek cognate with *tukhê*.

342. We often regain our health apparently by luck, especially when other reasons for doing so are not evident.

343. The Stoics distinguished *sunerga* causes as helping or intensifying, *sunaitia* as jointly causing the effect. Simplicius is here expressing a slightly different conception of a releasing cause. See notes 339 and 340.

344. Health.

345. As opposed to the eternal entities of the heavens.

346. All popular representations of the goddess.

347. The Greek word is *genesi* (dative plural of *genos*, which bears a variety of meanings including both 'nation' and 'pursuit'); if it means 'nations' here, Simplicius must have Constantinople in mind.

348. *Laws* 709B.

349. This Platonic diatribe seems to be a counterblast to Aristotle's relegation of luck to a subsidiary causal role. In the background is Plato's *Timaeus*, where the Errant Cause is portrayed as a plain fact of the cosmos; however much God or the Demiurge attempts to impose a rational pattern on the contents of the cosmos, he is constantly thwarted by the inherently unruly nature of matter. Here Simplicius gives us a glimpse of the Platonic-Plotinian scheme whereby an increasing movement away from the One comports an increasing element of luck.

350. i.e. the efficient cause.

351. i.e. the centre.

352. In the case of 'unchanging entities' such as mathematicals there is no place for efficient, final or material causal explanation – although this is challenged by Simplicius below.

353. Again Simplicius is putting forward a Platonist view to counter Alexander here. In the training of the philosopher in *Republic* 6 the soul passes from contemplation of sensible particulars to contemplation of the higher forms via the mathematicals. In particular the would-be philosopher, on leaving the Cave at

Republic 516A6, needs to acclimatise himself to the Intelligible World by contemplating the lower forms such as the mathematicals before proceeding to view the higher forms and ultimately the Good. Thus the mathematicals do serve a purpose and can be said to have a final cause.

354. Just as the mathematician *qua* mathematician will not look for final causes for mathematicals, the natural scientist *qua* natural scientist will not look for such final causes in the principles in natural science, since he is not concerned with theoretical philosophy.

355. Early natural scientists made the mistake, in their materialist reductionism, of imagining that the material description of anything provided a full causal explanation. I have taken the Greek word *eskhaton* ('last, ultimate') at 362,24 as referring to the last of the four causes mentioned in the *lemma*, but it is in fact the word used twice there by Aristotle in the sense of 'ultimate analysis'.

356. It is the nature of earth to sink, of fire to rise.

357. Simplicius brings in here one of the extra Platonist causes listed at 316,23-6.

358. *Timaeus* 31A.

359. Simplicius is showing the difference between a Platonist account (the Platonists are 'those who think that the Ideas (= Platonic forms) are causes etc.') and an Aristotelian account (as explained at 364,5ff.). 'These entities' refers to the particulars, and the phrase 'a single source' to the Platonic form. He perhaps has in mind the statement at *Republic* 10, 596A: 'we are in the habit of positing a single form in the case of the many particulars to which we attribute the same name'. For Aristotle's discussion of synonymity and homonymity see *Cat.* 1aff, and for his criticism of Platonic forms as causes see *Metaph.* 3, 996a27-30.

360. Perhaps *Metaph.* 12 (1069b25, 1071b21) rather than 1. The Greek numbers Λ (12) and A (1) are easily confused.

361. Simplicius' argument is: the formal and final causes of a natural entity are the same in form and number as it (and as each other), while the efficient cause is the same in form but not in substrate and number (cf. 264,5ff.); the perfected creature is one and the same as its final and formal causes in form and number, while it is the same in form but not in number and matter (substrate) as its efficient cause. How does nature fit into this scheme, since it has been seen earlier (ch. 1) both (a) as the changing cause of change, i.e. the efficient cause, and (b) as the form? Simplicius' answer is that nature *qua* efficient cause is the same in form, and that nature *qua* form is the same in number, as its effect. Nature is a special case. The discussion is limited to causes which lie within the remit of the natural scientist, such causes as nature which change while themselves being changed; hence the shift from 'primary origin of change' to 'all proximate efficient causes'.

362. Simplicius takes the subject of Aristotle's verb 'usually think of' as being the earlier natural scientists, whereas most modern editors take it as being less specific.

363. i.e. the remote and proximate efficient causes, the matter, the form.

364. Alexander's remarks seem to amount to no more than saying that there is a sizeable digression on the nature of the efficient cause. His subdivision is different from Simplicius' into unchanged and changed, the latter being further subdivided into remote and proximate rather than into perishable and imperishable. But in that the remote causes of all change in the sublunary world are the sun and the heavenly bodies, which are themselves changed but imperishable, while the proximate causes in the sublunary world are both themselves changed and perishable, the difference is narrow.

365. Simplicius picks up the reference to Aristotle's divine unmoved mover, which will be the subject of *Physics* Book 8.

366. Art and choice do not change *per se*, but they do alter and change their position *per accidens* because, for example, they move around with the agent of art and change when he moves. Similarly nature alters in a way within the entities of which it is the nature, which is why Simplicius has said above 'nature, or rather the things that cause change according to nature (for nature does not *per se* cause change, but with the help of the body in which it resides)'.

367. Simplicius draws a distinction between an efficient cause of being (*poiêtikon*) and a final cause of motion (*kinêtikon*). He is asking whether Aristotle's 'first of all things' could perform the creative role which the word *poiêtikon* implies. Aristotle's divine unmoved mover causes movement only as a final cause, but the late Neoplatonists construed it also as a creative or efficient (*poiêtikon*) cause of the world's being. Aristotle's efficient cause was so-called by the ancient commentators, rarely by Aristotle, and the word *poiêtikon* sometimes, as here, means creative as well as efficient. Simplicius claims that if Aristotle in this lemma is drawing a distinction among creative and efficient causes, as Alexander agrees, then Aristotle's God is indeed a creative and efficient cause. But if the distinction is only among causes of motion, then his God is being treated as a final cause. See Sorabji (1988) chs, 13 and 15.

368. 'The first of all things' performs a dual function as efficient and final cause.

369. Simplicius completes the discussion of 'the first of all things' as a cause of change by drawing a distinction between form 'taken just as it happens', which Alexander equates to mathematical form, and which precludes any sense of teleology, and form 'taken in a particular way because it is better so', which Alexander takes as form in natural entities, enmattered form, the formal aspect of the perfected entity towards which its nature has taken it. A thing's nature therefore is itself changed *per accidens* as it proceeds towards the end in view.

370. The argument is: straight is part of the definition in the case of a natural form such as man, as an end, in that the man as he develops aims at straightness and is potentially straight at any stage in the process; the straightness of geometrical form *qua* geometrical form is also potential, but not as an end.

371. A 'bottom upwards' as well as a 'top downwards' answer is necessary for a proper analysis. The former is mechanistic and is concerned with the matter, while the latter is teleological and concerns form, which here subsumes formal, final and efficient causes. As Simplicius goes on to point out, the question is fully discussed in *PA*. See further J.M. Cooper, 'Hypothetical necessity and natural teleology' in Gotthelf and Lennox (1987) 243-74.

372. Phaedo 97Bff.

373. The Greek word is *proêgoumenôs*. The idea of a principal cause is Stoic.

374. e.g. rotting the grain.

375. Does Aristotle think that rain is for the sake of the crops? Simplicius thinks so: 374,20. See D.J. Furley, 'The rainfall example in *Physics* 2.8' in Gotthelf (1985) 177-82.

376. DK 31B61.

377. This is the principle of natural selection. On how its use differs from modern uses of the principle, see Sorabji (1980) ch. 11.

378. This does not seem to be a direct quotation, but is the implication of *Phys.* 2.8, 199b26-8.

379. For the second syllogistic figure see note 253.

380. For the first syllogistic figure see note 45.

381. For the third syllogistic figure see note 45.

382. Here is the explicit interpretation, supported further by D.J. Furley, 'The rainfall example in *Physics* 2.8' in Gotthelf (1985) 177-82.

383. Diels suggests that this emendation is the work of Ambrosianus.

384. Alexander seems to be reading a variant text.

385. Presumably the natural scientists discussed earlier.

386. Whereas Aristotle denied both intellect (*nous*) and reason (*logos*) to animals (Sorabji (1993) chs, 12-16), late Neoplatonism is more nuanced. Proclus argued for animals possessing reason, and this might seem to follow from Plato's view that human souls can be reincarnated as animals. But this doctrine was ingeniously reinterpreted in various ways by other Neoplatonists (Sorabji (1993) 180-94).

387. Once again Simplicius takes soul to be opposed to nature, instead of being one kind of nature.

388. Lifeless (*apsuhka*) things.

389. See note 42.

390. cf. *Top*. 142b31.

391. cf. *GC* 4.3-4.

392. DK 31B62.

393. DK 31B61.

394. DK 31B62; KRS and Diels both translate *eideos* (shape) as 'warmth', an interpretation which seems generally favoured by scholars. See Wright (1981) 178.

395. Ross makes this last sentence a new point with a new paragraph.

396. For all words marked ‡ the Greek verb is *tukhein*.

397. Alexander's first premiss is effectively what Aristotle says at 199b15.

398. Simplicius points to some textual confusion. The Greek for 'wash' (*louesthai*) and 'ransom' (*luesthai*) are very similar, although 'left' (*apêlthe*) and 'released' (*aphêke*) are not. I have repunctuated since the Diels text offers no main verb.

399. There is a word apparently missing from the text here. I have supplied 'discussion' (*zêtêsin*) at the suggestion of Diels.

400. 'This' could refer either to matter or necessity.

401. See J.M. Cooper, 'Hypothetical necessity and natural teleology' in Gotthelf and Lennox (1987) 243-74.

402. For Aristotle's further discussion of hypothetical (or conditional) necessity see *PA* 1.1. All through this passage we see two types of necessity operating. Material necessity dictates a necessary result of material circumstances, e.g. something hard impacting on something soft will necessarily damage it. Hypothetical necessity lays down a condition, which is not a *per se* necessary circumstance, which dictates what materials are necessary, e.g. *if* a saw is to be effective it must be made of iron; but the converse is not a necessary truth, that if something is made of iron, it must be a saw.

403. *Timaeus* 68E.

404. *Phaedo* 99B.

405. Empedocles, not Leucippus: cf. *Cael*. 295a16ff.

406. Anaximenes, Anaxagoras and Democritus; cf. *Cael*. 294b14ff.

407. Simplicius misquotes Aristotle here by omitting 'in which ...'.

408. Simplicius seems to be confusing the text here. The word 'except' (*plên*) is added to the statement at 200a5-6 'the wall could not ...' in the previous sentence.

409. i.e. the materials – bricks, stone, wood etc.

410. Simplicius' teleological view is based on the Aristotelian maxim that what happens always or usually is for some purpose.

411. D.A. Russell suggests that a diagram was used by Aristotle: see Lacey (1993) 187n.877.

412. cf. *EE* 1227b32.

413. This passage (389,18-390,4) and the reprise (390,32-391,3) are difficult. For understanding Aristotle, it is essential to see that he has used the word 'because' and is discussing explanation, not mere entailment. In his usual manner, he hopes that a definition of straightness can be used as part of a deductive explanation of the interior angles of a triangle adding up to two right angles. His further point is that the explanation cannot be reversed. The character of two right angles does not help to explain the character of straightness, although it is a necessary condition of straightness having the character that it has. Simplicius ignores the point about explanation. The proof and explanation of the angles of a triangle adding up to two right angles which Aristotle knew was that given later by Euclid 1.32. It depends on a proof and explanation (Euclid 1.13) that the straight line is such that when one stands on another, it makes adjacent angles equal to two right angles. That in turn depends on understanding the nature of a straight line.

Simplicius' interpretation of Aristotle's words at 200a16ff. ('Because this is straight-sided, it is necessary that the internal angles of the triangle should add up to two right angles. But the converse is not a necessary truth – although if it were not true we would not have the straight') is:

(a) the internal angles of the triangle should add up to two right angles

(b) while the affirmation of the leading principle (this is straight-sided) does not necessarily follow from the affirmation of the consequence (the angles equal two right angles) ('But the converse is not a necessary truth')

(c) yet the denial of the leading principle (the denial of 'this is straight-sided') does necessarily follow from the denial of the consequence (the denial of 'the angles equal two right angles').

This sequence is followed clearly in the arithmetic example (389,28-31)

(a) 5 + 5 (leading principle affirmed) add up to 10 (consequence affirmed)

(b) 10 (consequence affirmed) is not necessarily 5 + 5 (leading principle) – it can be, e.g., 4 + 6

(c) but if no 10 (denial of consequence) then necessarily no 5 + 5 (denial of leading principle).

But his explanation of Aristotle's geometrical example is not so clear (as he admits at 390,32). It is not clear just what Aristotle means by 'the straight' at 200a17. He may be talking of straight lines in general, or of the particular straight line drawn by him in his classroom demonstration of the proof. Simplicius, as he makes clear at 390,35-6, considers that Aristotle is demonstrating a triangle. The leading principle in this case can be seen as 'this figure has straight sides'. His argument then runs:

(a) 389,23-5: this figure has straight sides (affirmation of leading principle) and the internal angles of a triangle add up to two right angles (affirmation of consequence)

(c) 389,25: if it does not have <its internal angles adding up to two right angles> (denial of consequence) then this figure does not have straight sides (denial of leading principle)

(b) 389,25-7: if the triangle has its three internal angles adding up to two right angles (affirmation of consequence) it is not necessary that this figure should have straight sides.

So far Simplicius has merely changed the order to (a) – (c) – (b). But then he makes the puzzling claim (389,27-8) that 'it is possible for the three <internal> angles of a four-sided figure, not just those of a triangle, to be equal to two right angles'. This is presumably parallel to the correct statement that it is possible for 4 + 6, 3 + 7 etc., not just 5 + 5, to add up to 10. But either Simplicius has made a mistake, since it is manifestly untrue that any three of the four internal angles of a plane four-sided figure add up to two right angles, or else he has switched to the field of solid geometry and is talking about, for example, the three internal angles at the apex of a four-sided pyramid (the Greek word *tetrapleuron* could refer to a solid as well as to a plane figure), or else he is referring to a curvilinear figure, where it is possible in terms of Greek geometry to construct such a figure (see Euclid book 1 *Definition* 8 (Heath (1920) 176-81) for non-straight angles.

At 390,32ff., after a note of explanation about external angles (390,33-5) he appears to repeat the (a) – (c) – (b) order.

(a) 390,35-391,1: this figure is straight-sided (affirmation of leading principle) so the three internal angles of a triangle add up to two right angles (affirmation of consequence)

(c) 391,1-2: if this is not so (denial of consequence) this figure is not straight-sided (denial of leading principle)

(b) 391,2-3: if the three internal angles add up to two right angles (affirmation of consequence) it is not necessary that this figure is straight-sided.

This involves (i) taking the Greek word *pantôs* in the sense 'necessarily', whereas its usual meaning in this work is 'always' (which would perhaps be a reference back to 389,27-8) – although it can also be translated 'entirely', and (ii) taking the Greek word *trigônon* (triangle) as shorthand for 'the straight and all the things we have said about it *vis-à-vis* the triangle and its angles'.

The final difficulty with this passage lies in the last sentence at 391,2. Diels prints the Greek letter *gamma* as if it were a numeral (*g* = 3). But everywhere else in the text, except in the passage at 389,29-31, where there is a sequence of numbers in the arithmetical demonstration and we would expect to find numerals rather than words, he prints the number as a word (*treis* = three). Also the number usually goes after the article, whereas here it precedes it. It might therefore be more sensible to print it as *g'* (= 'at any rate').

414. 10 can be made up 4 + 6 just as easily as 5 + 5.

415. The external angles of a triangle are formed by extending each side as follows:

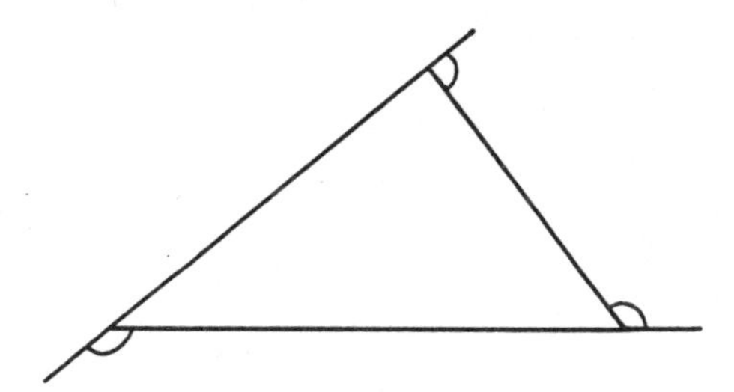

416. i.e. the material and the purpose.

417. *Timaeus* 47E. Plato is talking of the Divine Intellect as the Craftsman who shaped the physical universe and contrasting it with the wandering cause or the necessary constraints imposed by the materials the Craftsman had to work with.

418. Aristotle commonly uses the terms *suntheton* and *sunamphoteron* of the compound of form and matter.

419. *Metaph*. 1036a29ff.

420. Aristotle *DA* 1.1, 403a25-b9.

421. *ibid*.

Bibliography

* denotes volumes in this series.

C.M. Antonaccio, *An Archaeology of Ancestors: Tomb and Hero Cult in Early Greece*, Lanhan 1995

D.M. Balme (ed.), *The Human Embryo*, Exeter 1990

J. Barnes, M. Schofield and R. Sorabji (eds), *Articles on Aristotle 4*, London 1979

H.J. Blumenthal, 'Dunamis in Simplicius', in F. Romano and R. Loredana Cardullo (eds), *Dunamis nel Neoplatonismo*, Florence 1996

B.A. Brody (ed.), *Euthanasia and Suicide*, Boston 1989

W. Charlton, *Aristotle's Physics I,II*, Oxford 1970

J.J. Cleary and D.C. Shartin (eds), *Proceedings of the Boston Area Colloquium in Ancient Philosophy*, vol. 5 for 1988-9, Lanhan 1991

F.M. Cornford, *Plato's Cosmology*, London 1937

H. Diels, *Die Fragmente der Vorsokratiker*, 6th edition rev. by W. Kranz, Berlin 1952

L. Edelstein and I.G. Kidd (eds), *Posidonius: The Fragments*, vol. 2, Cambridge 1988

M.J. Edwards, *Philoponus: On Aristotle Physics 3*, London and Ithaca N.Y. 1994*

B. Fleet, *Plotinus: Ennead III.6*, Oxford 1995

M. Frede, 'The original notion of cause', in M. Schofield, M. Burnyeat, J. Barnes (eds) *Doubt and Dogmatism*, Oxford 1980

D.J. Furley, 'The rainfall example in *Physics* 2.8' in Gotthelf (1985) 177-82, reprinted in *Cosmic Problems*, Cambridge 1989.

A. Gotthelf (ed.), *Aristotle on Nature and Living Things*, Pittsburgh and Bristol 1985

A. Gotthelf and J.G. Lennox (eds), *Philosophical Issues in Aristotle's Biology*, Oxford 1987

H.B. Gottschalk, *Heraclides of Pontus*, Oxford 1980

M. Griffin, *Seneca: A Philosopher in Politics*, Oxford 1976

M. Griffin, 'Philosophy, Cato, and Roman Suicide', *Greece and Rome* 33, 1986, 64-77 and 192-202

W.K.C. Guthrie, *The Sophists* (vol. 5 of *History of Greek Philosophy*), Cambridge 1971

W.K.C. Guthrie, *Aristotle: An Encounter* (vol. 6 of *History of Greek Philosophy)*, Cambridge 1981

D. Harlfinger and M. Rashed, 'Survie byzantine du commentaire d'Alexandre d'Aphrodise la *Physique* d'Aristote. Vers une édition des scholies du Paris *sup. gr.* 643 et du *Paris gr.* 1859', *Philologus* forthcoming 1997

T.L. Heath, *Euclid Book 1*, Cambridge 1920

T.H. Irwin, *Aristotle's First Principles*, Oxford 1988

C.H. Kahn, *The Art and Thought of Heraclitus*, Cambridge 1979

B. Kendall and R.W. Thomson, *Definitions and Divisions of Philosophy by David the Invincible Philosopher*, Chico 1983

G.B. Kerferd, *The Sophistic Movement*, Cambridge 1981
O. Kern, *Orphicorum Fragmenta*, Berlin 1922
G.S. Kirk, J.E. Raven and M. Schofield (eds), *The Presocratic Philosophers* (2nd edition), Cambridge 1983
W. and M. Kneale, *The Development of Logic*, Oxford 1962
A.R. Lacey, *Philoponus: On Aristotle Physics 2*, London and Ithaca N.Y. 1993*
P. Lautner, 'Rival theories of self-awareness in late Neoplatonism', *Bulletin of the Institute of Classical Studies* n.s. 1, 1994, 107-16
H.G. Liddell, R. Scott and H.S. Jones, *A Greek-English Lexicon*, 9th edition, Oxford 1940
G.E.R. Lloyd, *Aristotle: The Growth and Structure of his Thought*, Cambridge 1968
D. Morrison, 'Philoponus and Simplicius on tekmeriodic proof', in E. Kessler (ed.), *Method and Order in Renaissance Philosophy of Literature: The Aristotle Commentary Tradition*, Dordrecht 1996
M. Rashed, 'Alexandre d'Aphrodise et la *Magna Quaestio*, Rôle et indépendence des scholies dans la tradition byzantine du corpes aristotélicien', *Les Études Classiques* 63, 1995, 295-351
M. Rashed, 'A new text of Alexander on the soul's motion?', in R. Sorabji (ed.), *Aristotle and After*, *Bulletin of the Institute of Classical Studies* supp. vol., 1996
W.D. Ross, *Aristotle's Physics*, Oxford 1936
W.D. Ross, *Aristotle: Fragmenta Selecta*, Oxford 1955
B. Snell, *Tragicorum Graecorum Fragmenta*, vol. 1, Göttingen 1986
R. Sorabji, *Necessity, Cause and Blame*, London and Ithaca N.Y. 1980
R. Sorabji, *Time, Creation and the Continuum*, London and Ithaca N.Y. 1983
R. Sorabji (ed.), *Philoponus and the Rejection of Aristotlelian Science*, London and Ithaca N.Y. 1987
R. Sorabji, *Matter, Space and Motion*, London and Ithaca N.Y. 1988
R. Sorabji (ed.), *Aristotle Transformed*, London and Ithaca N.Y. 1990
R. Sorabji, *Animal Minds and Human Morals: The Origins of the Western Debate*, London and Ithaca N.Y. 1993.
L. Spellman, *Substance and Separation in Aristotle*, Cambridge 1995
J.O. Urmson, *Simplicius: Corollaries on Place and Time*, London and Ithaca N.Y. 1992*
H. von Staden, *Herophilus: The Art of Medicine in Early Alexandria*, Cambridge 1989
F. Wehrli, *Die Schule des Aristoteles vol. 7: Herakleides Pontikos*, Basel and Stuttgart 1953
F. Wehrli, *Die Schule des Aristoteles vol. 8: Eudemos von Rhodos*, Basel and Stuttgart 1955
L.G. Westerink, *The Greek Commentators on Plato's Phaedo vol. 1: Olympiodorus*, Amsterdam 1976
P.H. Wicksteed and F.M. Cornford, *Aristotle Physics Books I-IV*, Cambridge Mass. and London (revised) 1957
C. Wildberg, *Philoponus: Against Aristotle on the Eternity of the World*, London and Ithaca N.Y. 1987*
C. Wildberg, *John Philoponus' Criticism of Aristotle's Theory of Aether*, Berlin 1988
M.R. Wright, *Empedocles: The Extant Fragments*, New Haven and London 1981

Appendix
The Commentators*

The 15,000 pages of the Ancient Greek Commentaries on Aristotle are the largest corpus of Ancient Greek philosophy that has not been translated into English or other European languages. The standard edition (*Commentaria in Aristotelem Graeca*, or *CAG*) was produced by Hermann Diels as general editor under the auspices of the Prussian Academy in Berlin. Arrangements have now been made to translate at least a large proportion of this corpus, along with some other Greek and Latin commentaries not included in the Berlin edition, and some closely related non-commentary works by the commentators.

The works are not just commentaries on Aristotle, although they are invaluable in that capacity too. One of the ways of doing philosophy between A.D. 200 and 600, when the most important items were produced, was by writing commentaries. The works therefore represent the thought of the Peripatetic and Neoplatonist schools, as well as expounding Aristotle. Furthermore, they embed fragments from all periods of Ancient Greek philosophical thought: this is how many of the Presocratic fragments were assembled, for example. Thus they provide a panorama of every period of Ancient Greek philosophy.

The philosophy of the period from A.D. 200 to 600 has not yet been intensively explored by philosophers in English-speaking countries, yet it is full of interest for physics, metaphysics, logic, psychology, ethics and religion. The contrast with the study of the Presocratics is striking. Initially the incomplete Presocratic fragments might well have seemed less promising, but their interest is now widely known, thanks to the philological and philosophical effort that has been concentrated upon them. The incomparably vaster corpus which preserved so many of those fragments offers at least as much interest, but is still relatively little known.

The commentaries represent a missing link in the history of philosophy: the Latin-speaking Middle Ages obtained their knowledge of Aristotle at least partly through the medium of the commentaries. Without an appreciation of this, mediaeval interpretations of Aristotle will not be understood. Again, the ancient commentaries are the unsuspected source of ideas which have been thought, wrongly, to originate in the later mediaeval period. It has been supposed, for example, that Bonaventure in the thirteenth century invented the ingenious arguments based on the concept of infinity which attempt to prove the Christian view that the universe had a beginning. In fact, Bonaventure is merely repeating arguments devised

* Reprinted from the Editor's General Introduction to the series in Christian Wildberg, *Philoponus Against Aristotle on the Eternity of the World*, London and Ithaca, N.Y., 1987.

by the commentator Philoponus 700 years earlier and preserved in the meantime by the Arabs. Bonaventure even uses Philoponus' original examples. Again, the introduction of impetus theory into dynamics, which has been called a scientific revolution, has been held to be an independent invention of the Latin West, even if it was earlier discovered by the Arabs or their predecessors. But recent work has traced a plausible route by which it could have passed from Philoponus, via the Arabs, to the West.

The new availability of the commentaries in the sixteenth century, thanks to printing and to fresh Latin translations, helped to fuel the Renaissance break from Aristotelian science. For the commentators record not only Aristotle's theories, but also rival ones, while Philoponus as a Christian devises rival theories of his own and accordingly is mentioned in Galileo's early works more frequently than Plato.[1]

It is not only for their philosophy that the works are of interest. Historians will find information about the history of schools, their methods of teaching and writing and the practices of an oral tradition.[2] Linguists will find the indexes and translations an aid for studying the development of word meanings, almost wholly uncharted in Liddell and Scott's *Lexicon*, and for checking shifts in grammatical usage.

Given the wide range of interests to which the volumes will appeal, the aim is to produce readable translations, and to avoid so far as possible presupposing any knowledge of Greek. Notes will explain points of meaning, give cross-references to other works, and suggest alternative interpretations of the text where the translator does not have a clear preference. The introduction to each volume will include an explanation why the work was chosen for translation: none will be chosen simply because it is there. Two of the Greek texts are currently being re-edited – those of Simplicius *in Physica* and *in de Caelo* – and new readings will be exploited by

1. See Fritz Zimmermann, 'Philoponus' impetus theory in the Arabic tradition'; Charles Schmitt, 'Philoponus' commentary on Aristotle's *Physics* in the sixteenth century', and Richard Sorabji, 'John Philoponus', in Richard Sorabji (ed.), *Philoponus and the Rejection of Aristotelian Science* (London and Ithaca, N.Y. 1987).

2. See e.g. Karl Praechter, 'Die griechischen Aristoteleskommentare', *Byzantinische Zeitschrift* 18 (1909), 516-38 (translated into English in R. Sorabji (ed.), *Aristotle Transformed: the ancient commentators and their influence* (London and Ithaca, N.Y. 1990); M. Plezia, *de Commentariis Isagogicis* (Cracow 1947); M. Richard, '*Apo Phônês*', *Byzantion* 20 (1950), 191-222; É. Evrard, *L'Ecole d'Olympiodore et la composition du commentaire à la physique de Jean Philopon*, Diss. (Liège 1957); L.G. Westerink, *Anonymous Prolegomena to Platonic Philosophy* (Amsterdam 1962) (new revised edition, translated into French, Collection Budé; part of the revised introduction, in English, is included in *Aristotle Transformed*); A.-J. Festugière, 'Modes de composition des commentaires de Proclus', *Museum Helveticum* 20 (1963), 77-100, repr. in his *Études* (1971), 551-74; P. Hadot, 'Les divisions des parties de la philosophie dans l'antiquité', *Museum Helveticum* 36 (1979), 201-23; I. Hadot, 'La division néoplatonicienne des écrits d'Aristote', in J. Wiesner (ed.), *Aristoteles Werk und Wirkung* (Paul Moraux gewidmet), vol. 2 (Berlin 1986); I. Hadot, 'Les introductions aux commentaires exégétiques chez les auteurs néoplatoniciens et les auteurs chrétiens', in M. Tardieu (ed.), *Les règles de l'interprétation* (Paris 1987), 99-119. These topics are treated, and a bibliography supplied, in *Aristotle Transformed*.

translators as they become available. Each volume will also contain a list of proposed emendations to the standard text. Indexes will be of more uniform extent as between volumes than is the case with the Berlin edition, and there will be three of them: an English-Greek glossary, a Greek-English index, and a subject index.

The commentaries fall into three main groups. The first group is by authors in the Aristotelian tradition up to the fourth century A.D. This includes the earliest extant commentary, that by Aspasius in the first half of the second century A.D. on the *Nicomachean Ethics*. The anonymous commentary on Books 2, 3, 4 and 5 of the *Nicomachean Ethics*, in *CAG* vol. 20, is derived from Adrastus, a generation later.[3] The commentaries by Alexander of Aphrodisias (appointed to his chair between A.D. 198 and 209) represent the fullest flowering of the Aristotelian tradition. To his successors Alexander was The Commentator *par excellence*. To give but one example (not from a commentary) of his skill at defending and elaborating Aristotle's views, one might refer to his defence of Aristotle's claim that space is finite against the objection that an edge of space is conceptually problematic.[4] Themistius (*fl.* late 340s to 384 or 385) saw himself as the inventor of paraphrase, wrongly thinking that the job of commentary was completed.[5] In fact, the Neoplatonists were to introduce new dimensions into commentary. Themistius' own relation to the Neoplatonist as opposed to the Aristotelian tradition is a matter of controversy,[6] but it would be agreed that his commentaries show far less bias than the full-blown Neoplatonist ones. They are also far more informative than the designation 'paraphrase' might suggest, and it has been estimated that Philoponus' *Physics* commentary draws silently on Themistius six hundred times.[7] The pseudo-Alexandrian commentary on *Metaphysics* 6-14, of unknown

3. Anthony Kenny, *The Aristotelian Ethics* (Oxford 1978), 37, n.3: Paul Moraux, *Der Aristotelismus bei den Griechen*, vol. 2 (Berlin 1984), 323-30.

4. Alexander, *Quaestiones* 3.12, discussed in my *Matter, Space and Motion* (London and Ithaca, N.Y. 1988). For Alexander see R.W. Sharples, 'Alexander of Aphrodisias: scholasticism and innovation', in W. Haase (ed.), *Aufstieg und Niedergang der römischen Welt*, part 2 *Principat*, vol. 36.2, *Philosophie und Wissenschaften* (1987).

5. Themistius *in An. Post.* 1,2-12. See H.J. Blumenthal, 'Photius on Themistius (Cod. 74): did Themistius write commentaries on Aristotle?', *Hermes* 107 (1979), 168-82.

6. For different views, see H.J. Blumenthal, 'Themistius, the last Peripatetic commentator on Aristotle?', in Glen W. Bowersock, Walter Burkert, Michael C.J. Putnam, *Arktouros*, Hellenic Studies Presented to Bernard M.W. Knox (Berlin and N.Y., 1979), 391-400; E.P. Mahoney, 'Themistius and the agent intellect in James of Viterbo and other thirteenth-century philosophers: (Saint Thomas Aquinas, Siger of Brabant and Henry Bate)', *Augustiniana* 23 (1973), 422-67, at 428-31; id., 'Neoplatonism, the Greek commentators and Renaissance Aristotelianism', in D.J. O'Meara (ed.), *Neoplatonism and Christian Thought* (Albany N.Y. 1982), 169-77 and 264-82, esp. n. 1, 264-6; Robert Todd, introduction to translation of Themistius *in DA* 3.4-8, in *Two Greek Aristotelian Commentators on the Intellect*, trans. Frederick M. Schroeder and Robert B. Todd (Toronto 1990).

7. H. Vitelli, *CAG* 17, p. 992, s.v. Themistius.

authorship, has been placed by some in the same group of commentaries as being earlier than the fifth century.[8]

By far the largest group of extant commentaries is that of the Neoplatonists up to the sixth century A.D. Nearly all the major Neoplatonists, apart from Plotinus (the founder of Neoplatonism), wrote commentaries on Aristotle, although those of Iamblichus (*c.* 250–*c.* 325) survive only in fragments, and those of three Athenians, Plutarchus (died 432), his pupil Proclus (410–485) and the Athenian Damascius (*c.* 462–after 538), are lost.[9] As a result of these losses, most of the extant Neoplatonist commentaries come from the late fifth and the sixth centuries and a good proportion from Alexandria. There are commentaries by Plotinus' disciple and editor Porphyry (232–309), by Iamblichus' pupil Dexippus (*c.* 330), by Proclus' teacher Syrianus (died *c.* 437), by Proclus' pupil Ammonius (435/445–517/526), by Ammonius' three pupils Philoponus (*c.* 490 to 570s), Simplicius (wrote after 532, probably after 538) and Asclepius (sixth century), by Ammonius' next but one successor Olympiodorus (495/505–after 565), by Elias (*fl.* 541?), by David (second half of the sixth century, or beginning of the seventh) and by Stephanus (took the chair in Constantinople *c.* 610). Further, a commentary on the *Nicomachean Ethics* has been ascribed to Heliodorus of Prusa, an unknown pre-fourteenth-century figure, and there is a commentary by Simplicius' colleague Priscian of Lydia on Aristotle's successor Theophrastus. Of these commentators some of the last were Christians (Philoponus, Elias, David and Stephanus), but they were Christians writing in the Neoplatonist tradition, as was also Boethius who produced a number of commentaries in Latin before his death in 525 or 526.

The third group comes from a much later period in Byzantium. The Berlin edition includes only three out of more than a dozen commentators described in Hunger's *Byzantinisches Handbuch*.[10] The two most important are Eustratius (1050/1060–*c.*1120), and Michael of Ephesus. It has been suggested that these two belong to a circle organised by the princess

8. The similarities to Syrianus (died *c.* 437) have suggested to some that it predates Syrianus (most recently Leonardo Tarán, review of Paul Moraux, *Der Aristotelismus*, vol.1 in *Gnomon* 46 (1981), 721-50 at 750), to others that it draws on him (most recently P. Thillet, in the Budé edition of Alexander *de Fato*, p. lvii). Praechter ascribed it to Michael of Ephesus (eleventh or twelfth century), in his review of *CAG* 22.2, in *Göttingische Gelehrte Anzeiger* 168 (1906), 861-907.

9. The Iamblichus fragments are collected in Greek by Bent Dalsgaard Larsen, *Jamblique de Chalcis, Exégète et Philosophe* (Aarhus 1972), vol. 2. Most are taken from Simplicius, and will accordingly be translated in due course. The evidence on Damascius' commentaries is given in L.G. Westerink, *The Greek Commentaries on Plato's Phaedo*, vol. 2, Damascius (Amsterdam 1977), 11-12; on Proclus' in L.G. Westerink, *Anonymous Prolegomena to Platonic Philosophy* (Amsterdam 1962), xii, n. 22; on Plutarchus' in H.M. Blumenthal, 'Neoplatonic elements in the de Anima commentaries', *Phronesis* 21 (1976), 75.

10. Herbert Hunger, *Die hochsprachliche profane Literatur der Byzantiner*, vol. 1 (= *Byzantinisches Handbuch*, part 5, vol. 1) (Munich 1978), 25-41. See also B.N. Tatakis, *La Philosophie Byzantine* (Paris 1949).

Anna Comnena in the twelfth century, and accordingly the completion of Michael's commentaries has been redated from 1040 to 1138.[11] His commentaries include areas where gaps had been left. Not all of these gap-fillers are extant, but we have commentaries on the neglected biological works, on the *Sophistici Elenchi*, and a small fragment of one on the *Politics*. The lost *Rhetoric* commentary had a few antecedents, but the *Rhetoric* too had been comparatively neglected. Another product of this period may have been the composite commentary on the *Nicomachean Ethics* (*CAG* 20) by various hands, including Eustratius and Michael, along with some earlier commentators, and an improvisation for Book 7. Whereas Michael follows Alexander and the conventional Aristotelian tradition, Eustratius' commentary introduces Platonist, Christian and anti-Islamic elements.[12]

The composite commentary was to be translated into Latin in the next century by Robert Grosseteste in England. But Latin translations of various logical commentaries were made from the Greek still earlier by James of Venice (*fl. c.* 1130), a contemporary of Michael of Ephesus, who may have known him in Constantinople. And later in that century other commentaries and works by commentators were being translated from Arabic versions by Gerard of Cremona (died 1187).[13] So the twelfth century resumed the transmission which had been interrupted at Boethius' death in the sixth century.

The Neoplatonist commentaries of the main group were initiated by Porphyry. His master Plotinus had discussed Aristotle, but in a very independent way, devoting three whole treatises (*Enneads* 6.1-3) to attacking Aristotle's classification of the things in the universe into categories. These categories took no account of Plato's world of Ideas, were inferior to Plato's classifications in the *Sophist* and could anyhow be collapsed, some

11. R. Browning, 'An unpublished funeral oration on Anna Comnena', *Proceedings of the Cambridge Philological Society* n.s. 8 (1962), 1-12, esp. 6-7.

12. R. Browning, op. cit. H.D.P. Mercken, *The Greek Commentaries of the Nicomachean Ethics of Aristotle in the Latin Translation of Grosseteste, Corpus Latinum Commentariorum in Aristotelem Graecorum* VI 1 (Leiden 1973), ch. 1, 'The compilation of Greek commentaries on Aristotle's Nicomachean Ethics'. Sten Ebbesen, 'Anonymi Aurelianensis I Commentarium in *Sophisticos Elenchos*', *Cahiers de l'Institut Moyen Age Grecque et Latin* 34 (1979), 'Boethius, Jacobus Veneticus, Michael Ephesius and "Alexander" ', pp. v-xiii; id., *Commentators and Commentaries on Aristotle's Sophistici Elenchi*, 3 parts, *Corpus Latinum Commentariorum in Aristotelem Graecorum*, vol. 7 (Leiden 1981); A. Preus, *Aristotle and Michael of Ephesus on the Movement and Progression of Animals* (Hildesheim 1981), introduction.

13. For Grosseteste, see Mercken as in n. 12. For James of Venice, see Ebbesen as in n. 12, and L. Minio-Paluello, 'Jacobus Veneticus Grecus', *Traditio* 8 (1952), 265-304; id., 'Giacomo Veneto e l'Aristotelismo Latino', in Pertusi (ed.), *Venezia e l'Oriente fra tardo Medioevo e Rinascimento* (Florence 1966), 53-74, both reprinted in his *Opuscula* (1972). For Gerard of Cremona, see M. Steinschneider, *Die europäischen Übersetzungen aus dem arabischen bis Mitte des 17. Jahrhunderts* (repr. Graz 1956); E. Gilson, *History of Christian Philosophy in the Middle Ages* (London 1955), 235-6 and more generally 181-246. For the translators in general, see Bernard G. Dod, 'Aristoteles Latinus', in N. Kretzmann, A. Kenny, J. Pinborg (eds), *The Cambridge History of Latin Medieval Philosophy* (Cambridge 1982).

of them into others. Porphyry replied that Aristotle's categories could apply perfectly well to the world of intelligibles and he took them as in general defensible.[14] He wrote two commentaries on the *Categories*, one lost, and an introduction to it, the *Isagôgê*, as well as commentaries, now lost, on a number of other Aristotelian works. This proved decisive in making Aristotle a necessary subject for Neoplatonist lectures and commentary. Proclus, who was an exceptionally quick student, is said to have taken two years over his Aristotle studies, which were called the Lesser Mysteries, and which preceded the Greater Mysteries of Plato.[15] By the time of Ammonius, the commentaries reflect a teaching curriculum which begins with Porphyry's *Isagôgê* and Aristotle's *Categories*, and is explicitly said to have as its final goal a (mystical) ascent to the supreme Neoplatonist deity, the One.[16] The curriculum would have progressed from Aristotle to Plato, and would have culminated in Plato's *Timaeus* and *Parmenides*. The latter was read as being about the One, and both works were established in this place in the curriculum at least by the time of Iamblichus, if not earlier.[17]

Before Porphyry, it had been undecided how far a Platonist should accept Aristotle's scheme of categories. But now the proposition began to gain force that there was a harmony between Plato and Aristotle on most things.[18] Not for the only time in the history of philosophy, a perfectly crazy proposition proved philosophically fruitful. The views of Plato and of Aristotle had both to be transmuted into a new Neoplatonist philosophy in order to exhibit the supposed harmony. Iamblichus denied that Aristotle contradicted Plato on the theory of Ideas.[19] This was too much for Syrianus and his pupil Proclus. While accepting harmony in many areas,[20] they could see that there was disagreement on this issue and also on the issue of whether God was causally responsible for the existence of the ordered

14. See P. Hadot, 'L'harmonie des philosophies de Plotin et d'Aristote selon Porphyre dans le commentaire de Dexippe sur les Catégories', in *Plotino e il neoplatonismo in Oriente e in Occidente* (Rome 1974), 31-47; A.C. Lloyd, 'Neoplatonic logic and Aristotelian logic', *Phronesis* 1 (1955-6), 58-79 and 146-60.

15. Marinus, *Life of Proclus* ch. 13, 157,41 (Boissonade).

16. The introductions to the *Isagôgê* by Ammonius, Elias and David, and to the *Categories* by Ammonius, Simplicius, Philoponus, Olympiodorus and Elias are discussed by L.G. Westerink, *Anonymous Prolegomena* and I. Hadot, 'Les Introductions', see n. 2 above.

17. Proclus in *Alcibiadem 1* p. 11 (Creuzer); Westerink, *Anonymous Prolegomena*, ch. 26, 12f. For the Neoplatonist curriculum see Westerink, Festugière, P. Hadot and I. Hadot in n. 2.

18. See e.g. P. Hadot (1974), as in n. 14 above; H.J. Blumenthal, 'Neoplatonic elements in the de Anima commentaries', *Phronesis* 21 (1976), 64-87; H.A. Davidson, 'The principle that a finite body can contain only finite power', in S. Stein and R. Loewe (eds), *Studies in Jewish Religious and Intellectual History presented to A. Altmann* (Alabama 1979), 75-92; Carlos Steel, 'Proclus et Aristotle', Proceedings of the Congrès Proclus held in Paris 1985, J. Pépin and H.D. Saffrey (eds), *Proclus, lecteur et interprète des anciens* (Paris 1987), 213-25; Koenraad Verrycken, *God en Wereld in de Wijsbegeerte van Ioannes Philoponus*, Ph.D. Diss. (Louvain 1985).

19. Iamblichus ap. Elian *in Cat.* 123,1-3.

20. Syrianus *in Metaph.* 80,4-7; Proclus *in Tim.* 1.6,21-7,16.

physical cosmos, which Aristotle denied. But even on these issues, Proclus' pupil Ammonius was to claim harmony, and, though the debate was not clear cut,[21] his claim was on the whole to prevail. Aristotle, he maintained, accepted Plato's Ideas,[22] at least in the form of principles (*logoi*) in the divine intellect, and these principles were in turn causally responsible for the beginningless existence of the physical universe. Ammonius wrote a whole book to show that Aristotle's God was thus an efficent cause, and though the book is lost, some of its principal arguments are preserved by Simplicius.[23] This tradition helped to make it possible for Aquinas to claim Aristotle's God as a Creator, albeit not in the sense of giving the universe a beginning, but in the sense of being causally responsible for its beginningless existence.[24] Thus what started as a desire to harmonise Aristotle with Plato finished by making Aristotle safe for Christianity. In Simplicius, who goes further than anyone,[25] it is a formally stated duty of the commentator to display the harmony of Plato and Aristotle in most things.[26] Philoponus, who with his independent mind had thought better of his earlier belief in harmony, is castigated by Simplicius for neglecting this duty.[27]

The idea of harmony was extended beyond Plato and Aristotle to Plato and the Presocratics. Plato's pupils Speusippus and Xenocrates saw Plato as being in the Pythagorean tradition.[28] From the third to first centuries B.C., pseudo-Pythagorean writings present Platonic and Aristotelian doctrines as if they were the ideas of Pythagoras and his pupils,[29] and these forgeries were later taken by the Neoplatonists as genuine. Plotinus saw the Presocratics as precursors of his own views,[30] but Iamblichus went far beyond him by writing ten volumes on Pythagorean philosophy.[31] Thereafter Proclus sought to unify the whole of Greek

21. Asclepius sometimes accepts Syranius' interpretation (*in Metaph.* 433,9-436,6); which is, however, qualified, since Syrianus thinks Aristotle is realy committed willy-nilly to much of Plato's view (*in Metaph.* 117,25-118,11; ap. Asclepium *in Metaph.* 433,16; 450,22); Philoponus repents of his early claim that Plato is not the target of Aristotle's attack, and accepts that Plato is rightly attacked for treating ideas as independent entities outside the divine Intellect (*in DA* 37,18-31; *in Phys.* 225,4-226,11; *contra Procl.* 26,24-32,13; *in An. Post.* 242,14-243,25).

22. Asclepius *in Metaph.* from the voice of (i.e. from the lectures of) Ammonius 69,17-21; 71,28; cf. Zacharias *Ammonius, Patrologia Graeca* vol. 85 col. 952 (Colonna).

23. Simplicius *in Phys.* 1361,11-1363,12. See H.A. Davidson; Carlos Steel; Koenraad Verrycken in n. 18 above.

24. See Richard Sorabji, *Matter, Space and Motion* (London and Ithaca, N.Y. 1988), ch. 15.

25. See e.g. H.J. Blumenthal in n. 18 above.

26. Simplicius *in Cat.* 7,23-32.

27. Simplicius *in Cael.* 84,11-14; 159,2-9. On Philoponus' *volte face* see n. 21 above.

28. See e.g. Walter Burkert, *Weisheit und Wissenschaft* (Nürnberg 1962), translated as *Lore and Science in Ancient Pythagoreanism* (Cambridge Mass. 1972), 83-96.

29. See Holger Thesleff, *An Introduction to the Pythagorean Writings of the Hellenistic Period* (Åbo 1961); Thomas Alexander Szlezák, *Pseudo-Archytas über die Kategorien*, Peripatoi vol. 4 (Berlin and New York 1972).

30. Plotinus e.g. 4.8.1; 5.1.8 (10-27); 5.1.9.

31. See Dominic O'Meara, *Pythagoras Revived: Mathematics and Philosophy in Late Antiquity* (Oxford 1989).

philosophy by presenting it as a continuous clarification of divine revelation[32] and Simplicius argued for the same general unity in order to rebut Christian charges of contradictions in pagan philosophy.[33]

Later Neoplatonist commentaries tend to reflect their origin in a teaching curriculum:[34] from the time of Philoponus, the discussion is often divided up into lectures, which are subdivided into studies of doctrine and of text. A general account of Aristotle's philosophy is prefixed to the *Categories* commentaries and divided, according to a formula of Proclus,[35] into ten questions. It is here that commentators explain the eventual purpose of studying Aristotle (ascent to the One) and state (if they do) the requirement of displaying the harmony of Plato and Aristotle. After the ten-point introduction to Aristotle, the *Categories* is given a six-point introduction, whose antecedents go back earlier than Neoplatonism, and which requires the commentator to find a unitary theme or scope (*skopos*) for the treatise. The arrangements for late commentaries on Plato are similar. Since the Plato commentaries form part of a single curriculum they should be studied alongside those on Aristotle. Here the situation is easier, not only because the extant corpus is very much smaller, but also because it has been comparatively well served by French and English translators.[36]

Given the theological motive of the curriculum and the pressure to harmonise Plato with Aristotle, it can be seen how these commentaries are a major source for Neoplatonist ideas. This in turn means that it is not safe to extract from them the fragments of the Presocratics, or of other authors, without making allowance for the Neoplatonist background against which the fragments were originally selected for discussion. For different reasons, analogous warnings apply to fragments preserved by the pre-Neoplatonist commentator Alexander.[37] It will be another advantage of the present translations that they will make it easier to check the distorting effect of a commentator's background.

Although the Neoplatonist commentators conflate the views of Aristotle with those of Neoplatonism, Philoponus alludes to a certain convention

32. See Christian Guérard, 'Parménide d'Elée selon les Néoplatoniciens', forthcoming.

33. Simplicius *in Phys.* 28,32-29,5; 640,12-18. Such thinkers as Epicurus and the Sceptics, however, were not subject to harmonisation.

34. See the literature in n. 2 above.

35. ap. Elian *in Cat.* 107,24-6.

36. English: Calcidius *in Tim.* (parts by van Winden; den Boeft); Iamblichus fragments (Dillon); Proclus *in Tim.* (Thomas Taylor); Proclus *in Parm.* (Dillon); Proclus *in Parm.*, end of 7th book, from the Latin (Klibansky, Labowsky, Anscombe); Proclus *in Alcib. 1* (O'Neill); Olympiodorus and Damascius *in Phaedonem* (Westerink); Damascius *in Philebum* (Westerink); *Anonymous Prolegomena to Platonic Philosophy* (Westerink). See also extracts in Thomas Taylor, *The Works of Plato*, 5 vols. (1804). French: Proclus *in Tim.* and *in Rempublicam* (Festugière); *in Parm.* (Chaignet); Anon. *in Parm* (P. Hadot); Damascius *in Parm.* (Chaignet).

37. For Alexander's treatment of the Stoics, see Robert B. Todd, *Alexander of Aphrodisias on Stoic Physics* (Leiden 1976), 24-9.

when he quotes Plutarchus expressing disapproval of Alexander for expounding his own philosophical doctrines in a commentary on Aristotle.[38] But this does not stop Philoponus from later inserting into his own commentaries on the *Physics* and *Meteorology* his arguments in favour of the Christian view of Creation. Of course, the commentators also wrote independent works of their own, in which their views are expressed independently of the exegesis of Aristotle. Some of these independent works will be included in the present series of translations.

The distorting Neoplatonist context does not prevent the commentaries from being incomparable guides to Aristotle. The introductions to Aristotle's philosophy insist that commentators must have a minutely detailed knowledge of the entire Aristotelian corpus, and this they certainly have. Commentators are also enjoined neither to accept nor reject what Aristotle says too readily, but to consider it in depth and without partiality. The commentaries draw one's attention to hundreds of phrases, sentences and ideas in Aristotle, which one could easily have passed over, however often one read him. The scholar who makes the right allowance for the distorting context will learn far more about Aristotle than he would be likely to on his own.

The relations of Neoplatonist commentators to the Christians were subtle. Porphyry wrote a treatise explicitly against the Christians in 15 books, but an order to burn it was issued in 448, and later Neoplatonists were more circumspect. Among the last commentators in the main group, we have noted several Christians. Of these the most important were Boethius and Philoponus. It was Boethius' programme to transmit Greek learning to Latin-speakers. By the time of his premature death by execution, he had provided Latin translations of Aristotle's logical works, together with commentaries in Latin but in the Neoplatonist style on Porphyry's *Isagôgê* and on Aristotle's *Categories* and *de Interpretatione*, and interpretations of the *Prior* and *Posterior Analytics*, *Topics* and *Sophistici Elenchi*. The interruption of his work meant that knowledge of Aristotle among Latin-speakers was confined for many centuries to the logical works. Philoponus is important both for his proofs of the Creation and for his progressive replacement of Aristotelian science with rival theories, which were taken up at first by the Arabs and came fully into their own in the West only in the sixteenth century.

Recent work has rejected the idea that in Alexandria the Neoplatonists compromised with Christian monotheism by collapsing the distinction between their two highest deities, the One and the Intellect. Simplicius (who left Alexandria for Athens) and the Alexandrians Ammonius and Asclepius appear to have acknowledged their beliefs quite openly, as later

38. Philoponus *in DA* 21,20-3.

did the Alexandrian Olympiodorus, despite the presence of Christian students in their classes.[39]

The teaching of Simplicius in Athens and that of the whole pagan Neoplatonist school there was stopped by the Christian Emperor Justinian in 529. This was the very year in which the Christian Philoponus in Alexandria issued his proofs of Creation against the earlier Athenian Neoplatonist Proclus. Archaeological evidence has been offered that, after their temporary stay in Ctesiphon (in present-day Iraq), the Athenian Neoplatonists did not return to their house in Athens, and further evidence has been offered that Simplicius went to Harrān (Carrhae), in present-day Turkey near the Iraq border.[40] Wherever he went, his commentaries are a treasurehouse of information about the preceding thousand years of Greek philosophy, information which he painstakingly recorded after the closure in Athens, and which would otherwise have been lost. He had every reason to feel bitter about Christianity, and in fact he sees it and Philoponus, its representative, as irreverent. They deny the divinity of the heavens and prefer the physical relics of dead martyrs.[41] His own commentaries by contrast culminate in devout prayers.

Two collections of articles by various hands have been published, to make the work of the commentators better known. The first is devoted to Philoponus;[42] the second is about the commentators in general, and goes into greater detail on some of the issues briefly mentioned here.[43]

39. For Simplicius, see I. Hadot, *Le Problème du Néoplatonisme Alexandrin: Hiéroclès et Simplicius* (Paris 1978); for Ammonius and Asclepius, Koenraad Verrycken, *God en wereld in de Wijsbegeerte van Ioannes Philoponus*, Ph.D. Diss. (Louvain 1985); for Olympiodorus, L.G. Westerink, *Anonymous Prolegomena to Platonic Philosophy* (Amsterdam 1962).

40. Alison Frantz, 'Pagan philosophers in Christian Athens', *Proceedings of the American Philosophical Society* 119 (1975), 29-38; M. Tardieu, 'Témoins orientaux du *Premier Alcibiade* à Harrān et à Nag 'Hammādi', *Journal Asiatique* 274 (1986); id., 'Les calendriers en usage à Harrān d'après les sources arabes et le commentaire de Simplicius à la *Physique* d'Aristote', in I. Hadot (ed.), *Simplicius, sa vie, son oeuvre, sa survie* (Berlin 1987), 40-57; id., *Coutumes nautiques mésopotamiennes chez Simplicius*, in preparation. The opposing view that Simplicius returned to Athens is most fully argued by Alan Cameron, 'The last day of the Academy at Athens', *Proceedings of the Cambridge Philological Society* 195, n.s. 15 (1969), 7-29.

41. Simplicius *in Cael.* 26,4-7; 70,16-18; 90,1-18; 370,29-371,4. See on his whole attitude Philippe Hoffmann, 'Simplicius' polemics', in Richard Sorabji (ed.), *Philoponus and the Rejection of Aristotelian Science* (London and Ithaca, N.Y. 1987).

42. Richard Sorabji (ed.), *Philoponus and the Rejection of Aristotelian Science* (London and Ithaca, N.Y. 1987).

43. Richard Sorabji (ed.), *Aristotle Transformed: the ancient commentators and their influence* (London and Ithaca, N.Y. 1990). The lists of texts and previous translations of the commentaries included in Wildberg, *Philoponus Against Aristotle on the Eternity of the World* (pp. 12ff.) are not included here. The list of translations should be augmented by: F.L.S. Bridgman, Heliodorus (?) in *Ethica Nicomachea*, London 1807.

I am grateful for comments to Henry Blumenthal, Victor Caston, I. Hadot, Paul Mercken, Alain Segonds, Robert Sharples, Robert Todd, L.G. Westerink and Christian Wildberg.

English-Greek Glossary

abridgement: *epitomê*
absence: *apousia*
abstract: *aphairein*
absurd: *atopos*
according to nature: *kata tên phusin*
accurate: *kurios*
achieve: *perainein*
act in advance: *proenergein*
action: *praxis*
affection: *pathos*
agreement: *sunthêkê*
in agreement: *sumpsêphos*
aim: *stokhazein*
aimless: *askopos*
alteration: *alloiôsis*
be altered: *alloiousthai*
ambiguous: *amphibolos*
analogy: *analogia*, *sunekdromê*
animal: *zôon*
be anxious: *spoudazein*
appropriate: *oikeios*
arbitrate: *diaitan*
argument: *epikheirêsis*, *epikheirêma*
arrangement: *taxis*
art: *tekhnê*
articulate: *diarthrein*
artificial: *tekhnêtos*
ask: *zêtein*
astromancy: *apotelesmatikê*
astronomer: *astrologos*
astronomy: *astrologia*
attempt: *epikheirein*
august: *polutimêtos*

bed: *klinê*
beginning: *arkhê*
being: *ousia*
belief: *dogma*, *doxa*
belong: *huparkhein*
blind: *tuphlos*
body: *sôma*
boundary: *peras*
broad outline: *kephalaion*
bubbling up: *anazesis*
bury: *katorussein*

causal explanation: *aitilogia*
cause: *aitia*, *aition*
caution: *eulabeia*
cessation of change: *stasis*
chance: *to automaton*
change (noun): *kinêsis*, *metabolê*
change (verb): *kinein*, *metaballein*
changed by something else: *heterokinêtos*
chapter: *kephalaion*
character: *kharaktêr*
characterize: *kharaktêrizein*
chilled: *apepsugmenos*
choice: *proäiresis*
cite: *paratithenai*
clear up: *anakathairein*
co-exist: *sunuphistasthai*
combine to produce: *sunistanai*
colour: *khrôma*
combined: *sumpeplegmenos*
come-to-be: *genesthai*, *gignesthai*
coming-to-be: *genesis*
commensurability: *summetria*
commentary: *exêgêsis*
common: *koinos*
comparable: *isostoikhos*
completion: *teleiotês*
completion, bring to: *sumplêroun*
composite: *sunamphoteron*, *suntheton*
compound: *sunamphoteron*, *suntheton*
concave: *koilos*
concavity: *koilotês*
concept: *ennoia*
concern oneself with: *pragmateuesthai*
consciousness, to have: *sunaisthesthai*
conclude: *epagein*, *sullogizein*
conclusion: *sumperasma*
concur: *sundramein*

concurrence: *sundromê*
condition: *diathesis*
configuration: *skhêmatismos*
conjunction: *sunapsis*
consequence: *akolouthia*
to be a consequence: *akolouthein*
consubstantiality: *sunupostasis*
containing: *sunektikon*
continuation: *sunokhê*
contradiction: *enstasis*
contradistinction: *antidiastolê*
contrast (noun): *parathesis*
contrast (verb): *antidiastellein*
contribute to: *suntelein*
contributory cause: *sunaition*
control: *dioikein*
convention: *nomos*
convex: *kurtos*
cooperating factor, to be a: *sunergein*
correspond: *analogein*, *antistrephein*
cosmogony: *kosmopoiia*
cosmos: *kosmos*
couch: *klinê*
crave: *khrêizein*
credible in itself: *autopistos*
cultured: *mousikos*
cure: *hugiazein*
current: *epipolazôn*

dance: *khoreia*
decay: *sêpedôn*
decrease: *meiôsis*
define: *horizein*
definition: *horismos*, *logos*
deliberate: *bouleuesthai*
deliberation: *boulê*
demonstrate: *apodeiknunai*
deny: *anairein*
depart from one's nature: *existasthai*
designation: *prosêgoria*
desire: *orexis*
destruction: *phthora*
determinate: *hôrismenos*
determined: *hôrismenos*
development: *diaplasis*
deviation: *rhopê*
differ: *diapherein*
difference: *diaphora*
difficult: *aporos*
difficulty: *aporia*, *enstasis*
dimension: *diastasis*
directive: *arkhitektonikos*
disavow: *apoginôskein*
discussion: *logos*
disposition: *diathesis*
distinction: *diakrisis*
distinguish: *diakrinein*, *diorizein*
distorted: *diastrophos*
distribution: *dianemêsis*
divine: *theios*
division: *diäiresis*, *merismos*
doctor: *iatros*
doctrine: *doxa*
drop: *apotithenai*
duality: *diploê*

eclipse: *ekleipsis*
effect: *aitiaton*
efficacy: *eukhrêstotês*
element: *stoikheion*
elemental: *stoikheiôdês*
embrace: *periekhein*
end in view: *skopos*, *telos*
enjoy substantial existence: *huphistasthai*
enmattered: *enulos*
enquire: *zêtein*
enquiry: *zêtêsis*
ensouled: *empsukhos*
entirety: *holotês*
eternal: *aiônios*
everlasting: *aïdios*
evident: *enargês*
examination: *ephodos*
example: *paradeigma*
existence: *hupostasis*
experience: *epeiresis*
explain: *didaskein*
explanation: *exêgêsis*, *paramuthia*
expound: *paradounai*
extension: *epitasis*

faint image: *parakhrôsis*
fall short of: *ektopizein*
fashion: *skhêmatizein*
figure: *skhêma*
final: *telikos*
first: *prôtos*
first actuality: *entelekheia*
fit: *summetros*
fitness: *euphuia*
flesh: *sarx*
follow: *akolouthein*
foresight: *pronoia*

formulation: *tupos*
fountainhead: *pêgê*
further division: *epairesis*
further examination: *epistasis*

geometrician: *geômetrês*
geometry of spheres: *sphairikê*
in general: *holôs*
generated: *genêtos*
generation: *apogennêsis*
goal: *skopos*, *telos*
god: *theos*
grow: *auxesthai*
growth: *auxêsis*, *blastêsis*, *dianastasis*, *ekphusis*

happen: *tunkhanein*
happiness: *eudaimonia*
harmony: *harmonia*
having extension: *diastatos*
having its being from itself: *authupostatos*
healing: *hugiansis*
health: *hugieia*
heating: *thermansis*
heavenly body: *astêr*, *ouranion*
heavens: *ouranos*
helmsman: *kubernêtês*
homoeomerous: *homoiomerês*
house: *oikia*
human: *anthrôpinos*
hypothesis: *hupothesis*

identity: *tautotês*
be ignorant: *agnoein*
image: *eikôn*
imagination: *phantasia*
imaging: *homoiôma*
immediate: *prosekhês*
imperishable: *aphthartos*
impotent: *adranês*
inappropriate: *akatallêlos*
include: *paralambanein*
indeterminate: *aoristos*
indicative: *deiktikos*
individual: *atomos*, *kath'hekaston*
induction: *epagôgê*
infer: *sêmeiousthai*
influence: *aphormê*
inherent: *emphutos*
be inherent: *enuparkhein*
inquiry: *methodos*
inseparable: *akhôristos*
instrument: *organon*
instrumental: *organikos*
intellect: *nous*
intelligible: *noeros*
intellective: *noeros*
intention: *dianoia*
interpret: *theôrein*
interpretation: *theôria*
interrupt: *diakoptein*
irrational: *alogos*
irregular: *ataktos*
irregularity: *anômalia*

join with: *koinônoun*
judge: *krinein*
just: *dikaios*

knowledge: *gnôsis*
knowledgeable: *gnôstikos*

lack: *apousia*
lacking in definition: *adioristos*
lacking in quality: *apoios*
leave: *apolimpanein*
length: *mêkos*
life: *zôê*
liken: *homoiousthai*
line: *grammê*
listing: *aparithmêsis*
long-lived: *polukhronios*
loss: *apobolê*
love: *philia*
luck: *tukhê*

be made: *genesthai*, *gignesthai*
make a distinction: *diastellein*
make-up: *sustasis*
match: *sunarmottein*
material: *hulikos*
mathematician: *mathêmatikos*
matter: *hulê*
mean: *sêmainein*, *theôrein*
means of judging: *kritêrion*
measure: *metron*
medical techniques: *iatrikê*
medicine: *iatreusis*
meet: *apantan*
meeting: *enteuxis*
menstrual fluid: *katamênion*
mind: *nous*
model: *paradeigma*

moon: *selênê*
motionless: *akinêtos*
movememt: *kinêsis*, *phora*
musical theory: *kanonikê*

natural: *phusikos*
natural scientist: *phusikos*, *phusiologos*
nature: *phusis*
not extended: *adiastatos*
notion: *ennoia*, *epibolê*
nourish: *trephein*
numinous: *daimonios*

objection: *enstasis*
oboe-playing: *aulêsis*
observation: *epistasis*
obscure: *aphanês*
observe: *historein*
obvious: *phaneros*
occur: *genesthai*, *gignesthai*
occurrence: *genesis*
offer a defence: *apologeisthai*
omit: *paratrekhein*
one in form: *monoeidês*
opinion: *doxa*, *doxasis*
opposite: *enantion*
optics: *optikê*
order: *dioikein*
organising: *kosmêtikos*
with organs: *organikos*
outcome: *apotelesma*
outright: *diarrhêdên*
outside calculation: *paralogos*
own: *oikeios*

pair: *duas*
paradigmatic (of cause): *paradeigmatikos*
part: *meros*
participate: *metekhein*
participation: *methexis*
particular: *idios*
per accidens: *kata sumbebêkos*
per se: *kath'auto*
perceive with the senses: *sunaisthesthai*
perception: *aisthêsis*
perfect realisation: *entelekheia*
perfection: *teleiotês*
Peripatetic: *Peripatêtikos*
perishable: *phthartos*
persistent: *monimos*
plane: *epipedon*
plant: *phuton*
Platonic: *Platônikos*
pleasure: *hêdonê*
point: *sêmeion*, *stigma*
popular: *dêmôdês*
portent: *teras*
position: *thesis*
positioning: *taxis*
positive: *kataskeuastikos*
posterior: *husteros*
potential: *dunamis*
power: *dunamis*
power of propagation: *gennêsis*
practical: *praktikos*
predetermine: *proörizein*
predicate: *katêgoria*
pre-exist: *proüparkhein*
premiss: *protasis*
presence: *parousia*
be present: *enuparkhein*
preservation: *sôtêria*
prevent: *empodizein*
prevision: *proörasis*
primary: *arkhêgikos*, *prôtos*
principal: *kurios*
principle: *arkhê* (cf. rational principle)
prior: *proteros*
be prior: *proêgeisthai*
privation: *sterêsis*
problem: *problêma*
procession: *proödos*
producing the form: *eidopoios*
production: *poiêsis*
productive: *gonimos*, *oistikos*, *poiêtikos*
proof: *epikheirêma*
propagate: *gennan*
propensity: *epitêdeiotês*
proper: *idios*, *oikeios*
proposition: *protasis*
protection: *asphaleia*
prove: *apodeiknunai*
proximate: *prosekhês*
pure: *eilikrinês*
to no purpose: *matên*

quality: *poiotês*
quantity: *posotês*
question: *problêma*

rational activity: *praxis*

rational principle: *logos*
reading: *lexis*
reason: *logos*
reasoning: *logismos*
reception: *metalêpsis*
reciprocal: *antistrophos*
regeneration: *palingenesia*
regular: *homalos*
regulate: *diakosmein*
regulative: *diakosmêtikos*
reproduce: *gennan*
resting on opinion: *endoxos*
resting on proof: *apodeiktikos*
result (noun): *apotelesma*
result (verb): *apantan*
reveal: *deiknunai*, *emphainein*
rising up: *dianastasis*
rule: *kanôn*

same in form: *homoeidês*
scholarly: *philoponos*
science: *epistêmê*
seed: *sperma*
self-evident: *gnôrimos*
self-moved: *autokinêtos*
separable: *khôristos*
separate (verb): *apokrinein*, *diakrinein*, *khôrizein*
separating: *diakritikos*
separation: *diakrisis*, *diastasis*, *ekkrisis*
shape: *morphê*, *skhêma*
shoot: *blastos*
short-lived: *oligokhronios*
show: *deiknunai*
signify: *sêmainein*
similarity: *homoiotês*
simple: *haplos*, *haploikos*
size: *megethos*
snub: *simos*
snubnosedness: *simotês*
soften: *paramutheisthai*
solid: *stereos*
soul: *psukhê*
source: *arkhê*
species: *diaphora*, *eidos*
spiritless: *apneumôn*
spherical: *sphairoeidês*
sprouting: *anablastêsis*
star: *astêr*
state: *hexis*
stop: *êremizein*
strict: *kurios*
strife: *neikos*
strip: *apoduein*
style: *lexis*
subject: *hupokeimenon*
subject matter: *problêma*
sublunary: *hupo selênên*
subsequent: *akolouthos*, *ephexês*
subsist: *huphistasthai*
subsist together with: *paruphistasthai*
substance: *hupostasis*
substrate: *hupokeimenon*
succession: *diadokhê*
sufficient: *autarkhês*
suitability: *epitêdeiotês*
sum up: *sumperainein*, *sunkephalaiousthai*
summary: *sunagôgê*
sun: *hêlios*
supervene: *epigignesthai*
surface: *epipolê*
syllogism: *sullogismos*
syllogistical: *sullogistikos*

term: *horos*
testimony: *marturia*
think: *nomizein*
thought: *dianoêsis*, *dianoia*
transfer: *metapherein*
transient: *euapoblêtos*
treat: *hugiazein*
treatise: *pragmateia*
treatment: *khrêsis*

unable to be grasped: *aperilêptos*
unchangeable: *atreptos*
unchanging: *ametablêtos*
unclear: *adêlos*
uncultured: *amousos*
understanding: *dianoia*, *gnôsis*, *katanoêsis*, *noein*
undisputed: *anamphilektos*
unembodied: *asômatos*
unformed: *arrhuthmistos*
unitary: *hênômenos*
uninterrrupted: *anekleiptos*
unity: *holotês*
unjust: *adikos*
universe: *kosmos*
unknowable: *agnôstos*
unlimited: *aoristos*
unmoved: *akinêtos*

unobvious: *aphanês*
usage: *tropos*
usefulness: *ôphelia*

vegetative: *phutikos*
vice: *kakia*
vice versa: enallax
virtue: *aretê*

warring: *makhê*
way: *tropos*
way of thinking: *ennoia*
weakness: *astheneia*
well-being: *euexia*
windfall: *periptôsis*
wish: *boulêsis*
with no purpose: *askopôs*
without limit: *apeiros*
witness: *martus*
work: *praxis*
world: *ouranos*

Greek-English Index

References are to the page and line numbers of the Greek text (H. Diels (ed.), *Commentaria in Aristotelem Graeca*, vol. 9, Berlin 1882), which appear in the margins of the translation. I have generally cited only two or three instances of each word. Verbs are given in the infinitive, nouns in the nominative case and adjectives in the nominative case (masculine), in the positive form even where the occurrence in the text is comparative or superlative.

apodeiknunai, to prove, demonstrate, 259,12.16
apodeiktikos, resting on proof, 328,20
apoduein, to strip, 295,14
apogennêsis, generation, 274,1
apoginôskein, to disavow, 278,18
apoios, lacking in quality, 320,28
apokrinein, to separate, 300,32
apolimpanein, to leave, 272,11
apologeisthai, to offer a defence, 260,21
aporia, problem, 342,27; 343,8
aporos, difficult, 330,21
apotelesma, result, outcome, 288,19.21; 342,8
apotelesmatikê, astromancy, 293,11
apotithenai, to drop, let go, 267,2
apousia, absence, lack, 259,16; 271,15; 280,15
apsukhos, lifeless, inanimate, 263,23; 346,10
aretê, virtue, 261,13
arkhê, beginning, principle, source *passim*
arkhêgikos, primary, 318,14
arkhitektonikos, directive, 304,22; 305,12
arrhuthmistos, unformed, 273,19.20
asaphês, obscure, 269,20
askopos, aimless, 302,2.5
askopôs, with no purpose, 335,13
asômatos, unembodied, 297,31.32
asphaleia, protection, 386,17
astêr, star, heavenly body, 290,22; 331,30
astheneia, weakness, 302,5
astrologia, astronomy, 290,9
astrologos, astronomer, 290,23
ataktos, irregular, 302,3; 331,25
atomos, individual, 298,5
atopos, absurd, 290,16; 292,32
atreptos, unchangeable, 320,23
aulêsis, oboe-playing, 261,15
autarkôs, sufficiently, 315,32
authupostatos, having its being from itself, 318,7
autokinêtos, self-moved, 298,22
autopistos, credible in itself, 272,15
auxesthai, to grow, 261,26
auxêsis, growth, increase, 261,28; 302,10
blastêsis, growth, 278,11
blastos, shoot, 274,3
boulê, deliberation, 385,17
boulêsis, desire, 268,15
bouleuesthai, to deliberate, 385,15

daimonios, numinous, 333,4.31; 341,15
deiknunai, to show, reveal, 259,10.19
deiktikos, indicative, 276,34
dêmôdês, popular, 295,12
diadokhê, succession, 311,18
diäiresis, division, 272,27; 300,6; 334,11
diaitan, to arbitrate, 269,32
diakoptein, to interrupt, 302,3
diakosmein, to regulate, 287,14
diakosmêtikos, regulative, 287,15
diakrinein, to separate, distinguish, 301,8; 334,35
diakrisis, distinction, separation, 260,4; 301,10
diakritikos, separating, 301,3
dianastasis, growth, rising up, 274,1; 289,34
dianemêsis, distribution, 260,36
dianoêsis, thought, 268,16
dianoia, thought, intention, 333,3; 336,25; 340,19; 345,24
diapherein, to differ, 260,18
diaphora, difference, species, 260,25; 261,10; 323,7.32
diaplasis, development, 302,16
diarrhêdên, outright, 332,35
diarthrein, to articulate, 333,32
diastasis, separation, dimension, 260,34; 290,15; 293,19
diastatos, having extension, 317,2
diastellein, to make a distinction, 279,17
diastrophos, distorted, 385,14
diathesis, disposition, 274,29; 310,5
didaskein, to explain, 259,25
dikaios, just, 261,14
dioikein, to control, order, 260,23; 332,5
diorizein, to distinguish, 260,8
diploê, duality, 298,22; 318,14
dogma, belief, 322,5
doxa, belief, opinion, doctrine, 259,8; 309,6
doxasis, opinion, 268,16

duas, a pair, 323,32
dunamis, potential, power, faculty, 262,23

eidopoios, producing the form, 276,1.6; 300,16
eidos, form *passim*
eikôn, image, 295,17.33; 324,28; 363,9
eilikrinês, pure, 287,23
ekkrisis, separation, 300,30
ekleipsis, eclipse, 291,29
ekphusis, growth, 278,36; 284,20
ektopizein, to fall short of, 381,1
emphainein, to reveal, 276,5
emphutos, inherent, 290,33
empodizein, to prevent, 288,26; 311,6
empsukhos, ensouled, 262,15; 294,18
enallax, vice versa, 276,13
enantion, opposite, 259,10; 260,34
enargês, evident, 260,20; 261,17
endoxos, resting on opinion, 328,20
ennoia, way of thinking, notion, conception, 270,16; 284,27; 330,31
enstasis, objection, contradiction, difficulty, 270,28; 307,19; 370,31; 380,19
entelekheia, first actuality, perfect realisation, 262,23; 277,32; 278,4
enteuxis, meeting, 357,8
enulos, enmattered, 294,6; 295,13
enuparkhein, to be present, inherent in, 259,7.13
epagein, to add, conclude *passim*
epagôgê, induction, 301,14
ep'elatton, less often the case, 261,22 *passim*
epereisis, experience, 299,26
ephexês, subsequently, in turn, 260,16
ephodos, examination, 370,13
epibolê, notion, 341,3
epidiairesis, further division, 335,12
epigignesthai, to supervene, 263,5; 286,36
epikheirein, to attempt, 276,4
epikheirêma, proof, argument, 278,35; 328,18
epikheirêsis, argument, 328,17
epipedon, plane, 290,5.31
epi pleiston, most often, 261,22
epi pleon, more often, covering a wider field, 270,36
epipolazôn, current, 357,4
epipolê, surface, 276,26
ep'isês, half the time, 334,17.18
epistasis, observation, further examination, 312,19; 363,33
epistêmê, science, 299,15
epitasis, extension, 303,28
epitêdeiotês, suitability, propensity, 280,17; 287,13; 314,14
epitomê, abridgement, 291,22
epi to polu, usually, 334,18; 338,13
êremizein, to stop, 287,11
euapoblêtos, transient, 274,29
eudaimonia, happiness, 321,28; 345,31; 346,1.2
euexia, well-being, 319,4
eukhrêstotês, efficacy, 373,26
eulabeia, caution, 293,28
euphuia, fitness, 287,15
exêgêsis, explanation, commentary, 270,22; 291,22
existasthai, to depart from one's nature, 310,4

genesis, coming-to-be, occurence, 259,10; 261,16
genesthai, come-to-be, occur, be made *passim*
genêtos, generated, 259,19
gennan, to reproduce, propagate, 261,26; 278,25
gennêsis, power of propagation, 278,20
geômetrês, geometrician, 276,14; 290,12
gignesthai, come-to-be, occur, be made *passim*
gnôrimos, self-evident, 272,5.6
gnôsis, knowledge, understanding, 259,13; 260,19; 333,4
gnôstikos, concerned with knowledge, cognisant, 301,31; 302,29; 313,29
gonimos, productive, 278,18
grammê, line, 290,32

haploikos, simple, 337,11
haplos, simple, 261,20; 262,24
harmonikê, harmony, 293,10; 294,27
hêdonê, pleasure, 322,5
hêlios, sun, 290,18
hênômenos, unitary, 301,9

heterokinêtos, changed by something else, 317,10.12
hexis, state, 274,29; 323,4
historein, to observe, 264,13
holotês, entirety, unity, 264,25; 336,34
holôs, in general, 260,16
homalos, regular, 292,27
homoeidês, the same in form, 297,3; 322,22; 364,22
homoiôma, imaging, 317,30
homoiomerês, homoeomerous, 273,23
homoiotês, similarity, 295,17.34
homoiousthai, to liken, 363,9
homônumos, bearing the same name but in a different sense, 269,13; 364,22
hôrismenos, determinate, determined, 261,21; 328,22
horismos, definition, 262,22; 269,11
horizein, to define, determine, 262,14; 263,6
hormê, impulse, influence, 261,27; 268,16
horos, term, 272,25.29
hôsautôs, in the same way, 334,13.26
hugieia, health, 279,16; 310,21
hugiansis, healing, 279,22.24
hugiazein, to cure, treat, 267,10; 279,16
hulê, matter *passim*
hulikos, material, 323,2
huparkhein, to belong, 262,22
huphistasthai, to enjoy (substantial) existence, subsist, 261,11; 289,5
hupo selênên, sublunary, 332,8; 334,15
hupokeimenon, substrate, subject *passim*
hupostasis, (substantial) existence, substance, 260,21; 287,6; 288,32
hupothesis, hypothesis, 275,33
husteros, posterior, 322,26

iatreusis, medicine, 279,13.15
iatrikê, medical techniques, 279,15
iatros, doctor, 260,24
idios, particular, proper, 274,8; 287,35
idiotês, particular characteristic, 298,29
isostoikhos, comparable, 354,9

kakia, vice, 261,14
kanôn, (general) rule, 293,30; 325,26
kanonikê, musical theory, 293,10
kanonizein, to make a general rule, 278,17
kata sumbebêkos, *per accidens passim*
kata tên phusin, according to nature *passim*
katamênion, menstrual fluid, 362,7; 391,26
katanoêsis, understanding, 287,6
kataskeuastikos, positive, 266,27
katêgoria, predicate, term, 265,13
kath'auto, *per se passim*
kath'hekaston, individual, 326,32
katorussein, to bury, 274,2
kephalaion, broad outline, chapter, 292,3; 388,11
kharaktêr, character, 289,5
kharaktêrizein, to characterize, 288,35
khoreia, dance, 331,30
khôristos, separable, separate, 291,2
khôrizein, to separate, 267,14.19; 291,3
khrêsis, treatment, 358,8
khrêizein, to crave, 270,33
khrôma, colour, 272,16; 276,27; 325,4
kinêsis, movement, change *passim*
klinê, bed, couch, 261,15
koilos, concave, 294,9
koilotês, concavity, 294,10
koinônoun, to join with, 276,1
koinos, common, 259,10; 260,10
kosmêtikos, organising, 301,3
kosmopoiia, cosmogony, 330,16
kosmos, cosmos, universe, 290,19; 331,18
krinein, to judge, 272,17
kritêrion, means of judging, 272,14
kubernêtês, helmsman, 268,7; 305,1
kubernêtikos, of a helmsman, 304,24
kurios, strict, accurate, principal *passim*
kurtos, convex, 294,8

lexis, style, reading, 261,4; 321,18; 329,14
logismos, reasoning, 341,13
logos, definition, discussion, reason *passim*

rational principle, 313,12.22.26.29.35.
loipon, finally, 260,12

makhê, warring, 379,34
marturia, testimony, 333,31; 335,2
martus, witness, 272,14
matên, to no purpose *passim*
mathêmatikos, mathematician, 260,24; 290,3
megethos, size, 276,27; 290,15; 291,12
meiôsis, decrease, 261,28
mêkos, length, 290,6
merismos, division, 327,22
meros, part, 261,19.29
metaballein, to change, 279,6
metabolê, change, 260,10; 274,31
metalêpsis, reception, 289,14
metapherein, to transfer, 298,29
metekhein, to participate, 297,1
methexis, participation, 296,34; 297,1
methodos, inquiry, 321,30
metron, measure, 264,17
monimos, persistent, 274,29
monoeidês, one in form, 291,14
morphê, shape, 275,5
mousikos, cultured, 260,11

neikos, strife, 300,16
nekros, lifeless, 287,19.21
noeros, intelligible, intellective, 296,27; 301,10; 391,32
nomizein, to think, 273,18
nomos, convention, 274,4
nous, mind, intellect, 261,12; 268,26; 272,24.28; 314,21.23; 317,17; 330,11
noein, understand, 272,19.22

oikeios, own, proper, appropriate, 259,7; 260,9
oikeiotês, appropriate relationship, 305,27
oikia, house, 261,15.31
oistikos, productive, 295,23
oligokhronios, short-lived, 332,12
ôphelia, usefulness, 390,13
optikê, optics, 294,26
orektikos, desiderative, 262,30,33;
orexis, desire, appetite, 261,13; 263,12; 379,30
organikos, with organs, 262,23; 268,20; 286,25
instrumental, 284,31; 315,18; 316,9.10.25; 318,20
organon, instrument, 315,15.17; 316,13
ouranion, heavenly body, 332,12
ouranos, world, heavens, 331,17
ousia, being, 262,15

palingenesia, regeneration, 379,22
paradeigma, example, model, 260,26; 295,18.34; 296,32; 310,32.35; 311,29; 312,1; 313,35; 314,15; 326,16; 363,2.4.5.7.9; 378,19
paradeigmatikos, paradigmatic (of cause), 298,17; 316,24; 317,31
paradounai, to expound, 259,4
parakhrôsis, faint image, 280,16
paralambanein, to include, 262,28; 263,14
paralogos, outside calculation, 340,19; 342,19
paramutheisthai, to soften, 270,28
paramuthia, explanation, 361,19
parathesis, contrast, 265,10
paratithenai, to cite, 273,10
paratrekhein, to omit, 270,23
parousia, presence, 278,2
paruphistasthai, to subsist together with, 353,35; 357,32
pathos, affection, 274,28; 298,20
pêgê, fountain-head, 268,30
perainein, to achieve, 329,26
Peripatêtikos, Peripatetic, 320,22
peras, boundary, 290,32.34
periekhein, to embrace, 322,30; 323,25
periptôsis, windfall, 261,16
phaneros, obvious, 272,8
phantasia, imagination, 379,30
phantazesthai, to imagine, 295,15
philia, love, 300,16
philoponos, scholarly, 291,21
phora, movement, 267,9; 271,4
phthartos, perishable, 259,19
phthora, destruction, passing out of being, 261,28
phusikos, natural, natural scientist, 259,9; 260,5
phusiologos, natural scientist, 309,8
phusis, nature *passim*
phutikos, vegetative, 262,16
phuton, plant, 261,19

Platônikos, Platonic, 320,23
poiêsis, production, 269,8; 301,10
poiotês, quality, 282,15; 291,30; 320,29
poiêtikos, productive, efficient, 259,5.23; 315,6; 337,28; 366,32; 367,27
polukhronios, long-lived, 332,13
polutimêtos, august, 317,17
posotês, quantity, 291,30; 292,8
pragmateia, treatise, 259,3; 326,29
pragmateuesthai, to concern oneself with, 290,18
praktikos, active, 261,12
praxis, action, work, rational activity, 261,14; 321,31; 345,28
proäiresis, choice, 261,13; 310,33
probouleuesthai, to make prior deliberation, 385,12.20
problêma, subject matter, question, object of enquiry, 261,3; 271,25; 333,35
proêgeisthai, to be prerequisite to, to be prior to, 260,19; 313,31
proenergein, to act in advance, 357,7
pronoia, foresight, 330,6
proödos, procession, 302,13
proörasis, prevision, 391,32
proörizein, to predetermine, 385,12
prosêgoria, designation, 320,19
prosekhês, proximate, immediate, 273,33; 305,8; 320,32
prosthêkê, addition, 267,5
proteros, prior, 322,26
prôtos, first, primary, 260,20; 262,6
proüparkhein, to pre-exist, be a prior condition, 260,33
protasis, proposition, premiss, 266,19; 320,7
psukhê, soul *passim*
psukhikos, of the soul, 261,27

rhopê, deviation, 358,7

sarx, flesh, 276,20
selênê, moon, 290,19
sêmainein, to mean, signify, 260,26
sêmeion, point, 290,32
sêmeiousthai, to infer, 307,9
sêpedôn, decay, 274,2
simos, snub, 294,9
simotês, snubnosedness, 294,10
skhêma, figure, shape, 266,5; 290,15
skhêmatismos, configuration, 368,21
skhêmatizein, to fashion, give shape to, 265,18; 291,9
skopos, end in view, goal, 357,13; 358,27
sôma, body, 261,20
sômatikos, bodily, 297,30
sôtêria, preservation, 379,34
sperma, seed, 320,33, 321,8
sphairikê, geometry of spheres, 293,9
sphairoeidês, spherical, 290,19
spoudazein, to be anxious, 272,7
stasis, cessation (of change/movement) *passim*
stereos, solid, 290,6
sterêsis, privation, 259,11.14
stigma, point, 290,6
stoikheiôdês, elemental, 259,4; 284,34; 312,5.9
stoikheion, element, 259,8; 309.5; 312,4; 320,3
letter (of alphabet), 275,28; 320,16
stokhastikos, requiring calculation, 385,19
stokhazein, to aim, 385,13
sullogismos, syllogism, 320,9
sullogistikôs, syllogistical, 266,5
sullogizein, to conclude, make inferences, reckon, syllogize, 264,8; 272,17; 343,27
summetria, commensurability, 310,21
summetros, fit, 272,32.34; 310,21
sumpeplegmenos, combined, 324,1.9
sumperainein, to sum up, 282,34; 288,13
sumperasma, conclusion, 266,18; 278,21; 320,8
sumplêroun, to bring to completion, 316,34-5
sumpsêphos, in agreement, 340,14
sunâidein, to be in keeping with, 270,16
sunagein, to infer, draw a conclusion, 266,5
sunagôgê, summary, 279,8
sunaisthesthai, to perceive with the senses, to have consciousness of, 272,20.35; 295,13
sunaition, joint cause, 316,25; 359,18
sunamphoteron, composite, compound, 264,31; 281,14; 294,6

Index of Works Cited in the Notes

Numbers in bold type refer to passages cited; numbers in ordinary type refer to the Notes. See also Subject Index for references to the translation. See also the following headings in the Subject Index: Alexander of Aphrodisias, Anaximenes, Antiphon, Eudemus, Geminus, Heraclides Pontus, Heraclitus, Menander, Parmenides, Posidonius, Porphyry, Protarchus, Syrianus, Thales, Themistius.

Subject Index

References are to the page and line numbers of the Greek text. These correspond to the numbers in the margin of the translation.